Heaven's net casts wide.
Though its meshes are coarse, nothing slips through.
—LAO-TSU

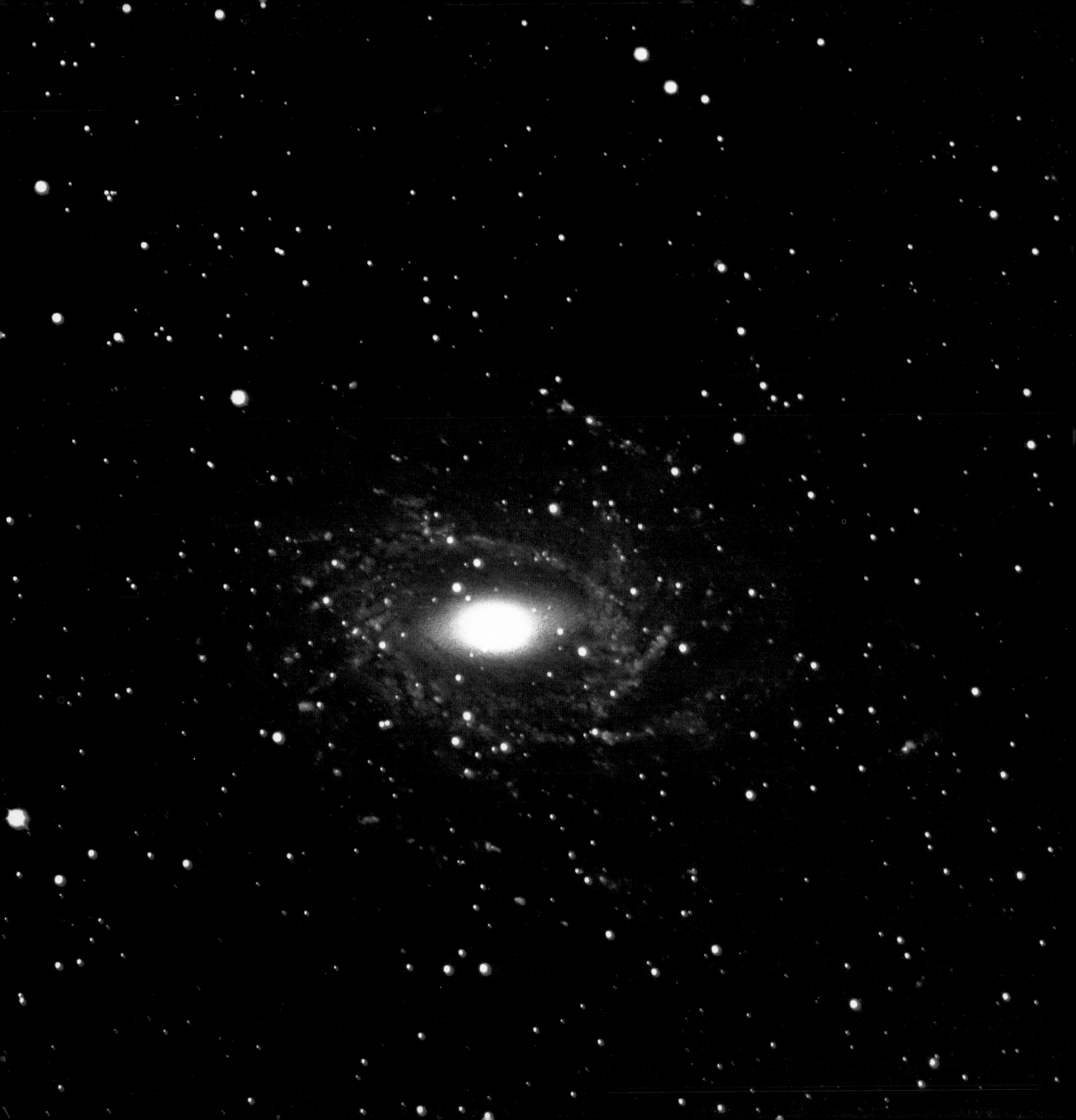

A Sierra Club Book

GALAXIES

Written and with photographs selected by TIMOTHY FERRIS

Illustrations by Sarah Landry

Stewart, Tabori & Chang, Publishers, New York

Frontispiece
The spiral galaxy designated NGC6744 is 300 million light-years distant from earth; a black-and-white photograph of this galaxy appears on page 109.

Distributed by Workman Publishing Company, Inc., 1 West 39th Street, New York, New York 10018.

Library of Congress Cataloging in Publication Data

Ferris, Timothy.
  Galaxies.

  Reprint. Originally published: San Francisco:
Sierra Club Books, c1980.
  Bibliography: p.
  Includes index.
  1. Galaxies.   I. Title.
QB857.F47 1982   523.1'12   81-21520
ISBN 0-941434-01-X   AACR2
ISBN 0-941434-02-8 (pbk.)

Printed and bound in Japan

Acknowledgments

The following are some of those who were kind enough to help out with *Galaxies*. Since none saw the entire book in its finished form, responsibility for errors or shortcomings remains solely that of the author.

The sun, the stars and seasons as they pass—some can gaze upon these with no strain of fear.

—HORACE

Information and Photographs

Halton Arp, Elly M. Berkhuijsen, Richard Berry, K. Alexander Brownlee, Lloyd Carter, S. Chandrasakhar, Mark R. Chartrand III, J. N. Clarke, James Cornell, A. G. de Bruyn, Terry Dickinson, Alan Dressler, Reginald Dufour, Vince Ford, Ken Franklin, Paul Gorenstein, J. Richard Gott III, Stephen T. Gottesman, John Graham, Edward J. Groth, B. W. Hadley, W. E. Harris, Eric B. Jensen, T. D. Kinman, Martha Liller, David Malin, Dennis Meredith, Simon Mitton, Richard Muller, Barry Newell, Rene Racine, Connie Rodriguez, D. H. Rogstad, Paul Routly, Ronald E. Royer, Vera Rubin, Allan Sandage, Jan Schafer, Malcolm Smith, Stephen Strom, Laird A. Thompson, Alar Toomre, Sindey van den Bergh, J. M. van der Hulst, Gerard de Vaucouleurs, Richard M. West, Fujiko Worrell, James D. Wray. Anglo-Australian Observatory, Astrophoto Laboratory, the Australian National University, Brooklyn College of the City University of New York, California Institute of Technology, Cambridge University, Carnegie Institution of Washington, the Cerro Tololo Inter-American Observatory, Dominion Astrophysical Observatory, European Southern Observatory, Griffith Observatory, Hale Observatories, Harvard College Observatory, Hayden Planetarium, Kitt Peak National Observatory, Lawrence Berkeley Laboratory of the University of California, Lick Observatory, Los Angeles Public Library, Max-Planck-Institut für Radioastronomie, McDonald Observatory, McMaster University, Mt. Stromlo and Siding Springs Observatories, New York Public Library, Princeton University, Rice University, Royal Observatory Edinburgh, Smithsonian Astrophysical Observatory, United States Naval Observatory, United States Naval Research Laboratory, Université de Montréal, the University of Chicago, the University of Florida, the University of Minnesota, the University of Texas at Austin, the University of Toronto, the Westerbork Radio Observatory.

Editorial Consultation

Timothy Crouse, Alan Dressler, J. Richard Gott III, Lynda Obst, Dennis Overbye, R. Bruce Partridge, Thomas M. Powers, Stephen Strom, Gerard de Vaucouleurs.

Advice and Encouragement

Jon Beckmann, Monica Brown, Jean B. Ferris, Wendy Goldwyn, Kathy Lowry, Lynda Obst, Bruce Porter, Thomas M. Powers, Delfina Rattazzi, Lisa Robinson, Allan Sandage, Alex Shoumatoff, Erica Spellman, Carolyn Zecca.

Illustrations

Sarah Landry

Research

Eileen Casey, Eustice Clarke, Juan de Jesus, Robert Ginsberg, Sandra Kitt, Judy Mitko, Terry Tammadge.

Dedication

To astronomers everywhere.

Table of Contents

Introduction

The Heavens . . . are now seen to resemble a luxuriant garden, which contains the greatest variety of productions, in different flourishing beds; and one advantage we may at least reap from it is that we can, as it were, extend the range of our experience to an immense duration. For, to continue the simile I have borrowed from the vegetable kingdom, is it not almost the same thing, whether we live successively to witness the germination, blooming, foliage, fecundity, fading, withering, and conception of a plant, or whether a vast number of specimens, selected from every change through which the plant passes in the course of its existence be brought at once to our view?

—WILLIAM HERSCHEL

Where Are We?

REBECCA: *I never told you about that letter Jane Crofut got from her minister when she was sick. He wrote Jane a letter and on the envelope the address was like this: It said: Jane Crofut; The Crofut Farm; Grover's Corners; Sutton County; New Hampshire; the United States of America.*

GEORGE: *What's funny about that?*

REBECCA: *But listen, it's not finished: the United States of America; Continent of North America; Western Hemisphere; the Earth; the Solar System; the Universe; the Mind of God—that's what it said on the envelope.*

GEORGE: *What do you know!*

REBECCA: *And the postman brought it just the same.*

—Thornton Wilder

Children are forever reinventing the game of the Long Address. Like all enduring games it is earnest at the root, and its concerns pass into adulthood. We should like to be able to write a flawless version of the Long Address, to dispel its elements of fancy and replace them with facts, as in adulthood we replace the potential with the actual, anticipation with experience, and ourselves with our children.

We are much closer to that goal now than we have ever been before. Having learned our way around our own planet, we have extended the scope of human vision deep into the cosmos, to find that our world is but one of many worlds in one of many galaxies. Having come to view the broad universe on a scale approaching that of its construction, we can attempt to recite the Long Address in earnest. In its present state, it reads something like this:

The Earth. A small planet orbiting the sun, a yellow dwarf star.

The Solar System. Nine known planets plus a variety of smaller bodies, among them comets and asteroids, all orbiting the sun. The earth, third planet out from the sun, follows an orbit with a radius of some ninety-three million miles; light from the sun takes 8.3 minutes to traverse this distance, so we might say that the earth is 8.3 light-minutes from the sun. The outermost of the known planets, Pluto, reaches a maximum distance of 4.6 billion miles, or a little under seven light-hours, from the sun. Beyond Pluto lies the realm of the comets; when they are taken into account, the radius of the solar system may amount to as much as a few light-days.

The Sun's Neighborhood. Within seventeen light-years of the sun are sixty known stars, the nearest of them 4.3 light-years away. The solar neighborhood has been described by the astronomers Peter van de Kamp and Sarah Lee Lippincott as resembling "some sixty small spheres—tennis balls, golf balls, marbles and a large proportion of smaller objects—spread at random through a spherical volume the size of earth." Most of these neighboring stars are utterly unspectacular. Despite their proximity, fewer than a dozen—Sirius, Alpha Centauri, Procyon, Altair among them—are bright enough to be seen with the unaided eye in the skies of earth. Most are dim bulbs like Barnard's Star, Wolf 359, and BD +36° 2147. Here in nearby space we find affirmed a lesson of nature already known to naturalists who study beetles or bacteria, that some of the least conspicuous inhabitants of creation are likely to be its most numerous.

Vicinity of the Orion Arm. The solar system lies near a luminescent spiral arm of our galaxy. As the stars of the constellation Orion lie within the arm at approximately its nearest point to us, less than two thousand light-years away, we have designated it the Orion Arm. The arm is not an object but a zone where new stars have been created and have lit up the interstellar gas surrounding them, like luminescent plankton churned up in the wake of a ship.

The Milky Way Galaxy. The starry congress to which the sun belongs is a major spiral galaxy, a giant wheeling system roughly one hundred thousand light-years in diameter and home to better than two hundred billion stars. This is quite an abundance of stars; if we were to launch expeditionary forces at such a fantastic rate that an expedition reached a new star in our galaxy every hour of the day and night, and we kept up this rate of exploration year after year, we would have visited fewer than half the stars in the Milky Way Galaxy in ten million years, five times longer than the present tenure of our species. So large and abundantly populated by stars is our galaxy that I doubt whether anyone would feel disappointed had it proved to be the whole of the cosmos. But it is only one among many galaxies.

The Local Group of Galaxies. The Local Group, a small cluster of galaxies bound together gravitationally, is dominated by a pair of large spirals, the Milky Way and the Andromeda galaxies. Its radius is roughly three million light-years. A map of the group appears on page 85.

The Local Supercluster. The Local Group lies near the outskirts of the Local Supercluster, a vast aggregation of clusters of galaxies, its radius perhaps one hundred million light-years. The Local Supercluster is discussed in Section Five of this book. Some of the groups belonging to it are mapped on page 153.

The Universe. The population of the universe has been estimated at one hundred billion major galaxies, a figure whose tidiness suggests its inexactitude. The universe is said to be expanding and evolving. By expanding it is meant that the clusters of galaxies are rushing apart from one another at velocities proportional to their distances; by interpolating the rate of expansion backward in time we can infer that all the stuff of the universe was once crammed together at a very high temperature and density. In short, the cosmos did not always look the way it does today. It has changed throughout its history of eighteen billion years or so, and continues to change today. Once highly homogeneous and uniform, the universe has differentiated into a startling variety of forms, among them galaxies and their stars, planets and ourselves. It is this that we mean when we say that the universe is evolving. Since expansion and evolution are functions of time, we may wish to add a temporal dimension to our version of the Long Address. When we have done so, it reads:

The Earth
The Solar System
The Sun's Neighborhood
Vicinity of the Orion Arm
The Milky Way Galaxy
The Local Group of Galaxies
The Local Supercluster
The Universe, circa eighteen billion years
 after the beginning of its expansion.

It was at this point that a species arose upon our planet which found itself able to discover the galaxies.

The Discovery of Galaxies

Knowing as we do today that the universe is amenable to investigation, that our telescopes can examine millions of galaxies at distances of millions of light-years, we may be tempted to feel impatient at our predecessors for having subscribed to cosmological doctrines that we now understand—thanks in part to their efforts—to have been inaccurate. The problems that they faced were not minor, as a brief review of the discovery of glaxies will suggest.

Though they shine with the light of many billions of suns, most galaxies are so distant that they look faint. Only three galaxies are visible to the naked eye from the surface of the earth. These are the two Magellanic Clouds, which lie in southern skies, and the Andromeda Galaxy, whose tenuous glow was aptly described by a seventeenth-century observer as resembling "the light of a candle shining through horn." Dozens more galaxies may be seen with a small telescope. But, again owing to their great distances, they cannot readily be resolved into their billions of constituent stars. Only when giant telescopes were wed with cameras and sophisticated equipment such as the spectroscope could the stars of distant galaxies be discerned and analyzed. It was for this reason that galaxies were not discovered—that is, were not understood to be what they are—until the twentieth century. Prior to that, the question of the large-scale structure of the universe was approached primarily by way of studying our own galaxy, as a city dweller might learn something of the nature of cities by studying solely the city in which he lives.

The Milky Way, the disk of our galaxy as seen from our perspective within the disk, is composed of a multitude of stars sufficiently far away that our eyes cannot resolve them. Instead, we see only a general wash of light. Observers as early as Democritus speculated that the Milky Way was made of stars, but not until Galileo trained a telescope on the skies could the hypothesis be confirmed. "The galaxy is nothing other than a mass of luminous stars gathered together," Galileo wrote. He told the news to Milton, who accordingly wrote in *Paradise Lost* of "The Galaxy, that Milky Way/Which nightly as a circling zone thou seest/ Powder'd with stars...."

Galileo's observations marked the beginning of the end of purely speculative cosmology. Prior to the invention of the telescope, when information about the stars was limited to what could be seen with the unaided eye, each theory of the structure of the universe had to be both concocted and judged almost entirely upon its merits as a creation of the imagination. Copernicus himself shared in this tradition, when on his deathbed he consented to publication of a cosmology that envisioned the earth orbiting the sun, for at the time the data on planetary motions fit the Copernican theory no better than they did the earth-centered cosmology of Ptolemy. Nor did the tradition of speculative cosmology die with the advent of Galileo. It continued at least until the eighteenth century, when it enjoyed a baroque efflorescence in the speculations of the young Immanuel Kant.

During the century between Galileo's death and Kant's boyhood several observers equipped with telescopes had noted the presence among the stars of ill-defined, luminous patches they called "nebulae." Today we know that several different sorts of objects were lumped together under the term. Most belonged to our galaxy; among them were clouds of gas illuminated by stars within them, shells of gas ejected by aging stars, a few indistinct clusters of stars. But some —the spiral nebulae—were galaxies in their own right, independent of the Milky Way.

It was Kant who guessed correctly that the spiral nebulae were galaxies. In 1755, when Kant was thirty years old, he wrote, "If a system of Fixed Stars which are related in their positions to a common plane, as we have delineated the Milky Way to be, be so far removed from us that the individual stars of which it consists are no longer sensibly distinguishable even by the telescope; if such a World of Fixed Stars is beheld at such an immense distance from the eye of the spectator situated outside of it, then this World will have the appearance of a small patch of space whose figure will be circular if its plane is presented directly to the eye, and elliptical if it is seen from the side or obliquely. The feebleness of its light, its shape, and the apparent size of its diameter will clearly distinguish such a phenomenon, when it presents itself, from all the stars that are seen as single." Word for word, it would be difficult to improve upon this description of the appearance of spiral galaxies.

Kant's cosmological speculations, published anonymously by a publisher who promptly went out of business, passed unnoticed at the time. Even had his theory won attention, there would have been at the time no way of putting it to an observational test. And that is the difference between the speculative cosmology of previous epochs and observational cosmology as practiced today. The vastly improved telescopes and other instruments now available to scientists permit them to test cosmological speculations by direct observation. The technological advances involved may be summarized in three categories—the use of the spectroscope to study the physics of stars, the development of precision telescopes able to measure the distances of nearby stars, and the building of giant telescopes with light-gathering power equal to the task of investigating remote galaxies.

The spectroscope enables researchers to examine the anatomy of light. The spectrum it produces might be compared to the musical score employed by a conductor to examine the parts played by individual musicians in an orchestra. The atoms of each element generate energy within a characteristic range of frequencies, as do the instruments in an orchestra, and within that range each atom plays a variety of melodies and harmonies from the study of which can be learned an enormous amount about the state of the atoms and of their environment. And if a star is moving toward or away from us, the frequency of the notes played by each of its atoms will be altered—shifted toward a higher frequency if the star is moving toward us, toward a lower frequency if it is moving away from us, in much the same way as a car horn sounds a different pitch if the car is approaching or receding. By analyzing this effect, known as the Doppler shift, astronomers can determine how rapidly a star is rotating, how fast it is moving in space, the extent of the churning motions within an interstellar cloud surrounding it, the velocity of stars in their orbits in another galaxy, and the velocity of galaxies as they participate in the general expansion of the universe. The uses of spectroscopy are almost limitless.

Just as the spectroscope helped to answer the ancient question of what the stars are made of, the equally fundamental question of the distances of stars was approached through the rise of precision astronomy in the nineteenth and early twentieth centuries, by way of the method of parallax. The fundamental principle of parallax is that the distance to a nearby star can be measured by changing our perspective on it. Hold your index finger at arm's length, close your left eye and sight the finger against a distant background: then close your right eye, open the left, and notice the apparent shift in the position of your finger; that is parallax. In astrometry—the precise measurement of the apparent positions of stars—the shift in perspective may be provided by taking two photographs of a relatively nearby star at an interval of six months, time enough to allow the earth in its orbit around the sun to shift its position by some one hundred eighty-six million miles, altering our perspective on the star the way switching from the right to the left eye altered our perspective on our finger. The diameter of the earth's orbit is not a very large distance by the standards of the stars, and astronomers working at the limits of this parallax method are obliged to measure shifts in perspective as minute as that which would be obtained by squinting alternately with one's right and left eyes at a finger six hundred miles away. But given an almost obsessive dedication to precision, reasonably accurate distances may be obtained for stars as far away as several hundred light-years.

Distances greater than a few hundred light-years are usually determined by estimating the instrinsic brightness of a star —known as its absolute magnitude—and comparing that value with its apparent brightness in the sky, known as apparent magnitude. The apparent magnitude of a star or other astronomical object decreases with the square of its distance, so determining its distance is simple once one knows how bright the star really is.

Astronomers have developed a variety of ingenious ways of estimating the absolute magnitudes of stars, most of them involving extrapolation from the studies of nearby stars whose distances could be determined by the parallax method. Nature has obliged them by creating variable stars of a sort whose period of brightness variation is directly related to their absolute magnitude. Once one knows the absolute magnitude of just a few such variables, one can use their cousins as distance indicators deep into the reaches of space. Particularly helpful in this regard have been the Cepheid variable stars, so bright that they can be identified in galaxies well beyond our own. Cepheids pulsate in brightness at intervals of as little as one day or as much as seventy days or more; the period of any given star betrays its absolute magnitude. Its distance can then be derived by comparing its absolute magnitude with its apparent magnitude. Distances to a number of neighboring galaxies, among them the giant spirals M81 and the Andromeda Galaxy, have been estimated by examining their Cepheid variable stars.

But Cepheid variables cannot be detected with existing telescopes at distances much greater than ten million light-years. To go farther, astronomers measure the size and brightness of glowing gas clouds and clusters of brilliant supergiant stars in galaxies, then estimate the distances of those galaxies by proceeding on the assumption that these denizens of other galaxies resemble their like here in the Milky Way. Explosions of stars as novae and supernovae also can be observed in distant galaxies and employed as a check against estimates made in other ways.

Beyond a few tens of millions of light-years the efficacy of these methods fades, and astronomers fall back upon the brightness and diameter of whole galaxies as indices of their distance. Here the assumption is that the dominant spiral in each cluster of galaxies will prove on the average to be comparable to that of dominant spirals like the Andromeda Galaxy in our Local Group.

Finally, the clusters of galaxies are rushing apart from one another as the universe expands; their velocities can be ascertained by measuring Doppler shifts in the spectra of their light. The farther away any given cluster of galaxies, the more rapidly it is receding in the expansion of the universe. (This is discussed more fully in Chapter Five.) So the distances of galaxies hundreds of millions or even billions of light-years away can be inferred from their velocity of recession.

That we can observe distant galaxies at all, much less determine their distances, analyze the compositions of their stars and interstellar clouds, measure their rates of rotation and from that estimate their mass, and otherwise chart them with a precision that few scientists a century ago would have thought possible—all this is due primarily to the advent of large telescopes. Unlike their colleagues in other physical sciences, astronomers cannot probe or dissect or experiment with the objects of their attention. Starlight washes down over the earth, a gentle rain indeed. All that the astronomer can do is to gather a little of it, bring it to a focus, and subject it to examination—by spectroscope, photometer, photographic plate or electronic detector. The more cosmic energy that can be gathered, the better, whether in the form of light, natural radio emanations, infrared and ultraviolet light, or the high-frequency energies of cosmic rays, X-rays and gamma rays. And more cosmic energy has been gathered and analyzed in our century than in all human history preceding it. This development more than any other has elevated our knowledge and awareness of the wider scheme of things.

At the vanguard of this scientific revolution was the construction by the American astronomer George Ellery Hale of large telescopes at the Mount Wilson and Palomar Observatories in California. It was at Mount Wilson that Harlow Shapley ascertained that the sun is located not near the center of the Milky Way Galaxy, as many had thought, but toward the outskirts of its disk. At Mount Wilson in 1924 Edwin Hubble determined that there are galaxies beyond ours, and that their stars are similar to stars found in our galaxy. Thus, questions concerning the existence of galaxies and our place within the Milky Way were transferred from the realm of speculation to the realm of verifiable fact.

From the work of Shapley, Hubble and their colleagues emerged two profound revelations about the nature of nature at large. The first is that the universe is far larger and more various than almost anyone had been prepared to believe. The other is that the breathtaking scope and diversity of the universe has arisen within the constraints of principles of nature—physical laws, as they are called—the same as those that pertain here at home. Nature everywhere plays by the same rules, and by learning those rules we can learn from nature, everywhere.

This second principle makes possible the science of astrophysics—the application of physical principles learned in laboratories here on earth to phenomena beyond the earth. We can estimate the mass of galaxies from their rotational velocities and from their interactions with one another because they obey the same principles—those elucidated by Newton and Einstein—as do falling apples on earth and the orbits of planets in the solar system. We can determine the composition of distant stars because they are made of the same sort of atoms as those we find here on earth or in the sun.

A third fundamental discovery of twentieth-century astronomy has been that of the expansion and what might be called the evolution of the universe. In 1929 Hubble, working in part from data supplied by a fellow astronomer, Vesto Slipher, discovered that remote galaxies are rushing apart from one another. Subsequent refinements in measurements of the rate of expansion have led to a modern estimate of roughly eighteen to twenty billion years since expansion began. Indirect confirmation of this time scale has come from the dating by astrophysicists of the oldest stars at some fifteen billion years of age, and from the discovery by radio astronomers of the cosmic background radiation, residual energy left over from the fiery moment when expansion began, its characteristics in accord with those of an expanding universe eighteen to twenty billion years of age.

Evolution, a word admittedly fraught with ambiguity, comes into play if we consider how the variety and diversity of the cosmos has increased with the passage of time. At the instant of the "big bang," any one scoopful of the stuff of the cosmos would have closely resembled any other—each would be essentially a batch of pure energy. Soon after expansion began much of the energy cooled to form the primordial elements hydrogen and helium, so that a token scoop would contain a greater variety of particles—hydrogen, helium and photons of energy—but each scoop would still generally resemble any other. Today the diversity of the cosmos is so

great that it is probably fair to say we have not begun to imagine it, much less to observe it. There are billions of galaxies, each with myriad varieties of stars and untold numbers of planets whose diversity of detail may perhaps be hinted at by the variety of life on earth and by our human thoughts about the cosmos. A scoopful of the cosmos drawn at random today might hold empty space, the alcohol molecules of an interstellar cloud, a dry ice snowball like those found on Mars, a rabbit's foot or words in a book. We see the universe today as a dynamic system within which human evolution plays a small but perhaps not discordant part. Herschel's vision of starry gardens where we find all sorts of plants at various stages in their lives has never seemed more nearly true.

About The Photographs

Writers who report on science solely in terms of its results are like hunters who shoot leopards solely for their skins. I am guilty of just such reporting in this book. I have presented the results of science, with little mention of the astronomers and astrophysicists whose work produced these results and made the book possible. I hope that they will not mistake this for ingratitude. My aim has been to encourage us to look at the galaxies directly, to appreciate that they are not solely specimens arrayed for scientific study, but that they are part of—most of—the natural world, as real and as worthy of our attention as are we who behold them. To do this I have been obliged to treat the scientific process as but a window onto the galaxies. I have tried to make that window so clear that we might occasionally forget that it is there.

The photographs in this book were taken, with a few exceptions, by astronomers employing some of the largest telescopes at various observatories around the world. The names of the observatories appear on page 5. There are many methods of depicting galaxies other than by photographs taken at optical wavelengths, but the human mind is strongly oriented around the visual sense, and so I have limited non-optical images to a few radio plots of galaxies and to the X-ray quasar image in Chapter Five.

The photographs are time exposures, obtained by aiming a telescope at a galaxy and exposing a photographic plate for as long as several hours while starlight seeps into the photographic emulsion. During this time a driving mechanism compensates for the earth's rotation and keeps the telescope trained on the galaxy, while the astronomer, or in some cases an automatic guiding system, makes minute corrections to compensate for refraction of light by the atmosphere or for inaccuracies in the driving mechanism.

The resulting photographs inevitably represent various compromises. Some compromise is involved in deciding upon exposure time. The interior sections of spiral galaxies contain a far higher density of stars and so are much brighter than the disks. A photograph that shows the spiral arms in detail must therefore overexpose the central region, while a photograph made to study the central regions will record little of the spiral arms. One can compensate for this by making two exposures, one for the arms and one for the central region, and combining them, but the result will produce an inaccurate impression of the galaxy's brightness profile. Another compromise involves the color sensitivity of the film chosen: a film preferentially sensitive to the red end of the spectrum will better record the ruddy bright nebulae that lie along the Milky Way, while a blue-sensitive film will suppress the clouds but bring out the young stars that lie within them. Astronomical photography, like all photography, contains an element of the impressionistic. At a few points I have endeavored to compensate for these limitations by presenting several photographs of the same galaxy taken in various wavelengths of light, as in the case of M82 (page 136), or by presenting detailed photographs of specific regions of a galaxy, as with the Andromeda spiral (pages 77-79).

The color photographs were in most cases produced by exposing three black and white plates, each limited by filters to one band of the spectrum, then combining them to produce a finished three-color print. The colors of galaxies cannot be seen directly by the human eye, even through the largest existing telescopes, for their light is too faint to stimulate the color receptors of the retina. And, since some elements of human judgment inevitably are involved in the production and reproduction of color photographs, slight differences in color balance may result when two observers create color images of the same galaxy. But the colors themselves are real, and the photographs represent the best efforts of astronomers to reproduce them accurately.

Galaxies and Human Thought

The study of galaxies by human beings has scarcely begun. Someone a century from now, reading what we thought about galaxies, would no doubt find much of it distorted, stunted or simply wrong. A map like the one that appears on page 42, which hazards to depict the sun's surroundings in our quarter of the Milky Way Galaxy, may by then seem as quaint as sixteenth-century maps of the New World, with their many errors and broad swaths of *terra incognita*. If our descendants smile at our ignorance, perhaps they will understand that we would have welcomed its banishment no less enthusiastically for its having been our own.

Looking back on our century they may be tolerant of us, keeping in mind that the discovery of galaxies and of the enormity and diversity of the cosmos came as something of a shock. We have lost the cosmos of our forebears, where the sky was wrapped around the world like a blanket and we walked upon a land that had been created for ourselves alone. It has not been easy for us to appreciate that there are many skies wrapped around many worlds and that the galaxies that harbor these worlds are real, that their fiery stars and planets and the ghostly clouds that waft among them are as much a part of the natural world as is any sun-basked meadow here on earth. It is not surprising that many of us feel nostalgia for older cosmologies, and that from this have sprung dissatisfied reactions to the new. One such reaction concerns location: Some feel remorse that we are removed from what we had imagined was a throne at the center of the universe. Another concerns diminution: How can we tiny creatures retain our self-esteem after having been exposed to the grandeur of the universe at large? A third concerns change: If the earth, stars and galaxies were born, are changing and will one day die, upon what rock may we build a church of any permanence?

We might be tempted to dismiss these objections by arguing that it doesn't matter how we feel about the galaxies, that their existence is a fact, and that we should accept that fact inasmuch as the secret of learning is to love the truth. But the existence of these feelings is fact, too, and most of us share in them to some degree. Stolid indeed is the student of galaxies who has felt no sense whatever of dislocation, disenfranchisement or vertigo at the sight of them. So those of us who feel that the benefits of the discovery of galaxies outweigh any trauma it may have produced might do well to explain why we feel that way.

As to location: It is true that we do not occupy the center of the universe. It seems that there *is* no center of the universe —except perhaps in the highly technical sense of a null point that might be discernible in the flux of the cosmic background radiation—and if such a center were located, one can scarcely imagine any reason why we would want to live there. Nor is it clear that any distinction ought to have been bestowed upon our imaginary occupancy of the center of things in the first place; the center of the world of Christian cosmology, for example, was occupied not by God or the angels but by Satan.

The universe, we now realize, is highly equitable in terms of location. The view is splendid from almost everywhere. If the sun were located in another galaxy we could just as easily observe the universe at large, and take photographs of galaxies such as the ones that appear in this book, upon one page of which would appear the Milky Way Galaxy. Here we are in the solar system, squired on the arm of a magnificent spiral galaxy; surely this is not grounds for complaint.

As to dimension: It is true that we are tiny relative to the cosmos. *Everything* is tiny relative to the cosmos—even a galaxy is but one among billions—and to fret about this is to confuse size with stature. We are well advised to bow to no tyranny of mere size, to heed the lesson of Lao-tsu, Aristotle, Leonardo and Darwin, who teach that the truth is less often attained by gaping at the grand than by scrutinizing the small. The human body is a galaxy to a microbe, yet without microbes the body would not live an hour. If we feel awe, let us address it not to dimension but to being—and we share being with the galaxies themselves.

As to change: The stars, once thought of as symbols of constancy, are seething fireballs hurtling along in a changing galaxy in a changing universe, and this realization clearly has its unsettling side. But so did the illusion of a changeless cosmos that preceded it. When the heavens were thought to be eternal and unalterable, it was tempting to regard the turmoil of human affairs as fundamentally different in character from the workings of nature. To live on earth was to be condemned to mutability and corruption, while the stars revelled in immutability and incorruptibility. The fundamental error of imagining that human life was fundamentally different from the rest of nature gave rise to many an odd doctrine. Aristotle argued that only the heavens were pure, Plato that the real world is static and archetypical and the flux of perceived existence but an illusion. But, then as now, others were willing to conceive of themselves as immersed in a changing cosmos. If Plato in his Olympian home is today disturbed by the cosmos in flux we believe we have discovered, his predecessor Heraclitus, who saw change in everything, warms his hands on the fires of the galaxies.

The reassuring aspect of the portrait of the universe we now see drawn across the sky lies in its reconciliation of humanity with the material world. That we are part of our galaxy is literally true. The atoms of which we are formed were gathered together in the toilings of a galaxy, their fantastical assembly into living creatures was nourished by the warmth of a star in a galaxy, we look at the galaxies with a galaxy's eye. To understand this is to give voice to the silent stars. Stand under the stars and say what you like to them. Praise or blame them, question them, pray to them, wish upon them. The universe will not answer. But it will have spoken.

T.F.

1/The Milky Way:
A Spiral Galaxy Viewed from Within

We too once lived in this house of stars....

—MURMURS OF EARTH

A Journey to the Center of the Milky Way

...Thoughts,
Which ten times faster glide than the sun's beams....
—SHAKESPEARE

Our sun and its planets lie in the environs of the Milky Way Galaxy. To go to the center of the galaxy would require navigating a distance of some thirty thousand light-years. Such a journey lies far beyond the technological capacity of our species in this century; interstellar distances are vast, the energy required to traverse them enormous. Some say we shall never be able to do it. Others say we might. No one expects that we shall do it soon.

Yet we can make the trip today, aboard the ship of the imagination. This may seem like mere daydreaming, but dreams have preceded earlier journeys, as when our forebears contemplated oceanic horizons in the days before we mastered the seas. And the sights to be seen during such a journey need not be pure imagination; we have learned enough about the galaxies to predict in general terms what we might hope to see if we traveled through them.

If we need further encouragement, let us consider the remarkable accommodation of science to fantasy represented by the time-dilation effect in Einstein's special theory of relativity. The theory, verified in many an experiment, tells us that the passage of time slows down dramatically aboard a spaceship that is accelerated to velocities approaching that of light. (The speed of light itself can never be attained, it adds.) We can spend energy to buy time.

Owing to this effect a starship able to maintain an acceler-ation equal to the force of gravity here on earth could reach the center of our galaxy, thirty thousand light-years distant, in under twenty-five years of on-board time. Neighboring galaxies could be achieved in less than thirty years, and gulfs separating clusters of galaxies crossed in perhaps a decade longer. So let us imagine ourselves aboard such a ship, and see where it might take us.

The ship's appointments may be left to the preference of each passenger. We might envision a giant vessel complete with baseball teams, string quartets, a hardwood copse and trout pond, and a crew of thousands selected from back-grounds sufficiently varied to insure that things will never quite run smoothly. Or a more modest cruise starship, with a tiny nightclub, an indefatigable recreation director, many outside cabins with portholes. Or a military starship, all drums, boots and salutes. Each to his own. There is ample room in the imagination for imaginary starships, as there is in the cosmos for real ones.

The day of our departure is sad, its farewells permanent. We travelers will be able to take advantage of the time-dilation effect. Friends and families who stay at home will not. They will have been dead for tens of thousands of years by the time we reach the center of our galaxy. Together we sing the anthem of interstellar explorers everywhere, a song of final farewell. Then we depart.

The early years of the voyage pass uneventfully while our ship accumulates velocity. Years pass before we can celebrate having attained the distance of the nearest extrasolar star, Alpha Centauri, a little over four light-years out. The sun is now but a dot of light in the constellation Taurus. Soon it will become difficult to identify the dim little sun in the sky.

In the years that follow, the stars crawl across the sky, slowly distorting the constellations we have known on earth until most are unrecognizable. Our course takes us along the plane of our galaxy directly toward its center. Our view is composed of stars and of the clouds of dust and gas that lie in the spaces between the stars. The interstellar clouds are mostly dark, but when we encounter one of the spiral arms of our galaxy, we find them lined with bright nebulae—regions where newly formed stars have lit up the surrounding clouds—and the sight of these glowing shoals cheers us as we speed on.

Many a flower of these starry pastures could hold our attention—the high density dwarf stars, neutron stars and black holes, the endlessly varied matchups of multiple stars, the variable and flare stars, and the billions of ordinary stars like our sun, not to mention their planets. But we must hurry on.

After decades of travel, the interstellar clouds at last fall away. Ahead lies the central region of the galaxy, an elliptical cosmos of stars glowing through the relatively unsullied spaces with fantastic clarity. The color of this great egg is that of a bloodied yolk, the red and orange light of old stars that have been burning steadily for billions of years. Behind us the inner portions of the dusty disk hang like the walls of a canyon; thousands of light-years down one wall we can discern the elbow joint where one spiral arm emerges from the central regions and begins a winding path that will eventually take it out past our sun.

We plunge into the central regions. Country dwellers on our first visit downtown, we are surprised at the congestion of the stars, their hustling pace. Their light is as warm in hue as torch light. They pursue jitterbug orbits that seem hurried by the standards of the solar region, and the clearances among them are narrow. Yet all conduct their affairs without colliding.

Our destination is the nucleus of the galaxy. We can see its brilliant lamp ahead. What will we find there? An enormous star cluster, sitting like a clutch of diamonds at the center of the galactic diadem? The ominous warren of a black hole, a creature out of the Inferno rather than the Paradiso?

It is just at this point of the voyage of our imagination that we must turn away. We know that our galaxy has a nucleus, but do not yet know enough about it to be able to describe it. The captain orders our course altered, and our ship describes a sweeping arc that takes it up and out of the plane of the galaxy. Ahead lies intergalactic space.

Figure 1

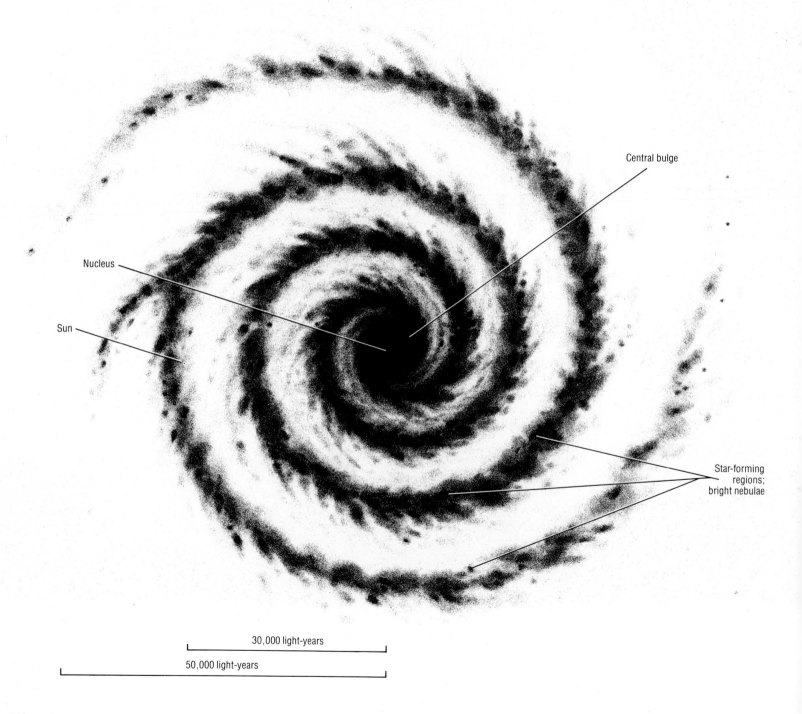

Central bulge

Nucleus

Sun

Star-forming
regions;
bright nebulae

30,000 light-years

50,000 light-years

Figure 2

Figure 1, 2. Milky Way Plan and Side Views
The anatomy of a normal spiral galaxy is presented in these views of
the Milky Way. The galaxy is centered on a compact nucleus surrounded
by a roughly spherical realm of stars usually referred to as the central
bulge. A spherical halo made up of scattered older stars embraces the
entire galaxy; here are to be found many globular clusters. Most of the
interstellar gas and dust of the galaxy, as well as most of its stars,
occupy the flattened disk. The spiral arms represent portions of the disk
made evident by the hosts of brightly shining new stars recently formed
there. The spiral arms are depicted schematically, as our galaxy has not
yet been well mapped beyond the solar neighborhood.

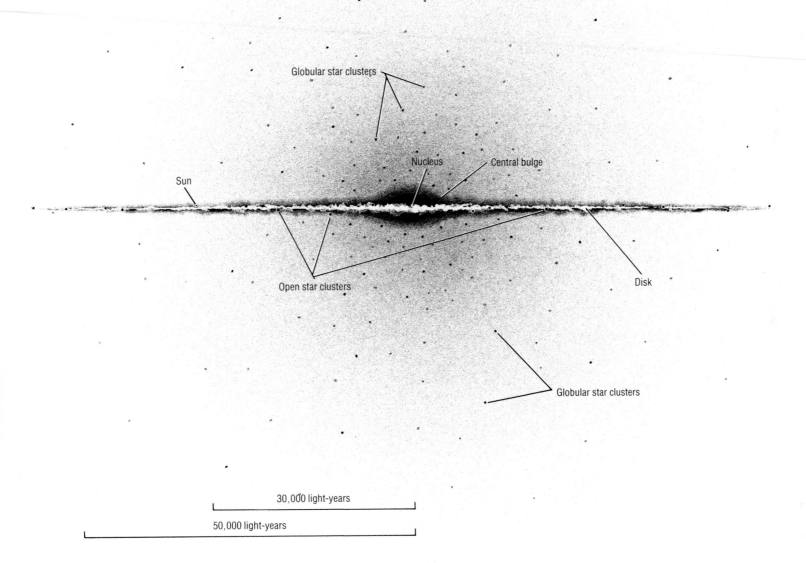

Globular star clusters

Nucleus   Central bulge

Sun

Open star clusters

Disk

Globular star clusters

30,000 light-years

50,000 light-years

find themselves with nowhere to sit and so depart. In this fashion, the sun in its core converts nearly five million tons of matter into energy each second. The energy that has lost the game of musical chairs makes its way slowly—this takes millions of years—to the surface, where it is radiated into space. It is a matter of perspective whether we care to think that stars employ nuclear fusion to release energy, or that they fuse atomic nuclei and produce energy as a byproduct.

Many stars vary in brightness over periods of days or months, some wildly so. Fortunately for we who depend upon its even temper, the sun is at most only mildly variable. On the basis of historical and climatological records, the sun seems to have altered its energy output somewhat in the long term. In the short term, its most spectacular signs of distemper take the form of periodic eruptions on the surface that hurl solar material out into space.

The photograph (page 21) shows one unusually large eruption from the surface of the sun. Solar material, primarily hydrogen gas—in the cosmos one is forever encountering hydrogen gas—is being ejected like a bubble bursting from the surface of a whitewater brook. The granular appearance of the solar disk results from the action of convection currents, hot towers of upwelling material alternating with descending cells of cooler material.

The photograph was made in ultraviolet light. The energy emitted by the sun comes in a variety of wavelengths, including ultraviolent and infrared light; it even manages some output in the very long wavelength radiation we call radio. But it radiates most strongly along the portion of the electromagnetic spectrum we call visible light. This is no coincidence; we evolved on a planet bathed in sunlight, and our eyes evolved as to make the most efficient use of its energy.

Stars and Interstellar Space

Nature plays conjurer's tricks, producing endless diversity from the most ordinary of ingredients, pulling from its top hat not only rabbits and bouquets and conjurers themselves, but stars. Stars are made from the simplest of ingredients, hydrogen gas (and hydrogen is the simplest of the elements) mixed with some helium (the second simplest element), and traces of more complex atoms. A congealing gas cloud becomes a star once it has compacted itself to a sufficiently high density that the heat generated in the crush of its core rises to a point high enough to fuse atomic nuclei and release energy. Gravitational force, the universal attraction of matter

for matter, tends to collapse the star. Energy bubbling up from its core tends to push it apart. The balance between these two forces maintains a star's composure. A star goes through various changes as it ages, yet these changes are dictated almost entirely by how much mass it started out with; for instance, very massive stars burn more rapidly and age more quickly than stars of average mass.

Stars are remarkable in their variety. There are stars smaller than earth and stars larger than earth's orbit, stars younger than human civilization and stars nearly as old as the universe, stars harder than diamond and gasbag stars so diffuse that much of them is thinner than air, hot blue stars and dim stars that glow the ruby hue of a cooling coal, variable stars that pulse like jellyfish, flare stars that brighten as suddenly as a campfire doused with kerosene, single stars like the sun, and double stars, triple and quadruple stars.

The stellar population of our galaxy is estimated at something over two hundred billion. The photograph (right) shows a few of them. Their apparent crowding is an illusion created by the fact that we are seeing thousands of light-years into the depths of space; stars that appear to be stacked almost next to one another actually are separated by many light-years. Stars generally have plenty of room; a few dozen baseballs scattered across all North America would crowd one another in comparison with the leeway enjoyed by an average star in an average galaxy.

The stars we see in our galaxy also vary widely in age. Some are extremely old, some very young; most, like our sun, lie between those extremes. It follows that we ought to find evidence of stellar birth and death going on around us, just as in the human community we find that some persons are younger, some older, and that births and deaths are constantly taking place. And considerable evidence has accumulated that such is the case. In our galaxy may be found new stars being born and old stars in their death throes. The sites of some of these events appear on the following pages.

The spaces between the stars of our galaxy are littered with clouds of dust and gas. Stars form from such clouds, which are characteristically very thin, typically thinner than a

---

2 The Milky Way in its tangle of stars and interstellar clouds offers us a firsthand look at how a spiral galaxy—in this case, our own—appears from a vantage within and toward one side of the galactic disk.

laboratory vacuum, but so vast that they have enough mass in the aggregate to make billions of suns.

At any given time most of the atoms in an average interstellar cloud drift along on their own. Some couple with other atoms to form molecules, and molecules in turn wander the wastelands of space. For stars to form from such a cloud, enough of these wandering atoms must be brought together so that gravity, a very weak force, is able to tether them and arrest their independent meanderings. Once this has happened, the bundle of atoms that results is able to capture other atoms that it encounters, binding them to the group, slowly increasing the group's mass and with it its gravitational attraction. Star seeds like these are growing today in many places in our galaxy, each scarcely noticeable yet awesome in its potential, like an embryonic cell in a womb.

Interstellar clouds are generally dark and inconspicuous except when illuminated by stars that have just formed there, or where silhouetted against a background of stars. The two clouds pictured here are seen in silhouette.

Many different sorts of terminology can be applied to interstellar clouds. Most convenient is to lump them under the term "nebula," from the Latin for "mist," or "cloud." Illuminated interstellar clouds are called bright nebulae. Unilluminated clouds are called dark nebulae.

4

3 The dark interstellar cloud known as the Coalsack, nearly six hundred light-years away and seventy light-years in diameter, is nestled against the foot of the Southern Cross, which lies on its side in this photograph (left). In reality, only Beta Crux and Delta Crux, the stars at the ends of the short arm of the cross, occupy the same celestial neighborhood as does the Coalsack. Alpha Crux, the bright star at the foot of the cross, and Gamma Crux, the star at its head, are both in the foreground, at respective distances of about three hundred seventy and two hundred twenty light-years.

4 The Cone Nebula, twenty-six hundred light-years distant, is part of an extended interstellar cloud. The bright stars in the background apparently formed out of the cloud recently in cosmic history; their light now silhouettes the foreground part of the cloud.

Stars Being Born

THE ORION NEBULA

The first stars to form within a condensing cloud bestow the gift of light upon their progenitor. Previously dark, the cloud now bursts forth with a bouquet of color that makes bright nebulae among the most arresting sights in the sky. Some of the illumination consists of light from the young stars reflected by dust grains in the surrounding cloud. But most of it is produced when gas in the cloud, far more abundant than dust, is ionized—electrically charged—by starlight, and glows by reradiating the received energy, as does the gas in a neon light.

Bright nebulae like these are found lining the arms of spiral galaxies, where density waves set up by the rotation of the galaxy have recently passed through, promoting the condensation of interstellar clouds into stars. The sun currently finds itself near one of the arms of our galaxy, and as a result, we are presented with an excellent view of the bright nebulae that adorn the arm. These include the Eta Carina and Rosette nebulae and, closest at hand, the Orion Nebula. The Trifid, Lagoon and Eagle nebulae belong to another spiral arm nearer the center of the galaxy than our own.

The stars that illuminate the Orion Nebula are celestial infants, some of them less than five hundred thousand years old. One is estimated to have begun shining only about two thousand years ago; still embedded in the dark cloud from which it formed, it is invisible in these photographs but can be detected at the wavelengths of infrared light, which penetrates the dust and gas as visible light cannot.

---

5 The Horsehead Nebula is part of a large dark cloud of which the Orion Nebula forms a bright spot. Here, a little less than one hundred light years from the Orion Nebula, part of the enormous dark cloud may be seen silhouetted against a slightly more distant part of the cloud whose sheets of gas have been excited by starlight until they glow. The Horsehead itself is an eddy, a slowly spinning ball of gas that may be expected eventually to condense into new stars. Swirling at velocities approaching fifteen miles per second, it will one day no longer resemble a horsehead, having transformed itself like an earthly cloud on a summer day.

6 (overleaf left) Astronomers studying the Orion Nebula (= M42 = NGC1976) have found evidence of infant stars there, still wrapped in swaddling clothes of gas and dust (page 30).

7 (overleaf right) The interior of the Orion Nebula, known as the Trapezium, glows with the delicate green hue of ionized oxygen. Although the nebula is composed primarily of hydrogen gas, molecules of oxygen and formaldehyde have been found scattered through it, as have many others including those of carbon monoxide, hydrogen cyanide, ammonia, water and methyl alcohol. It is not yet understood just how the atoms of these thin clouds managed to link up into such complicated molecules.

THE EAGLE NEBULA

If we could speed up our sense of time until thousands of years were speeding by in the wink of an eye, we would see bright nebulae like these burst into light, deliver themselves of a shower of stars, then fade back into darkness. As it is, we see each nebula frozen at a stage in the process. The light that sets the nebula aglow comes from bright young stars that recently formed within it. The dark congealings of gas and dust, particularly prominent in the Eagle Nebula, are on their way to forming more stars.

The study of what might be called stellar embryology is still in its early stages, and a great deal remains to be learned about how galaxies make stars. But the story can be recounted at least tentatively in general terms.

The interstellar clouds arrayed through the vast tracts of space in a galaxy like ours abide most of the time in a state of passivity. They toil sluggishly along in space, responsive to the ghostly promptings of the galaxy's magnetic and gravitational fields. Once in a long while a star happens by, gobbling up a swath of the cloud as it goes. Nothing much else happens. At any given time much of the interstellar medium is a heavenly Dead Sea.

But through this sea pass waves. Density waves resulting from resonances generated by the gravitational interaction of the stars of the galaxy propagate across the galactic disk in a spiral pattern. When a density wave passes through an interstellar cloud, its effect is to compress the cloud. If the cloud is terribly thin, the passage of the wave will have only a temporary effect, fleeting as the stirring of dead leaves in a breeze. But where the interstellar cloud was sufficiently dense to start with, the wave can compress it until the cloud's gravitational field becomes strong enough to begin to draw the cloud still more tightly together. Once begun, this process will tend to continue. The cloud forms knots and eddies of ever-increasing density that draw more of the surrounding gas and dust in upon themselves, growing to become globules like the Horsehead Nebula and those that we see in the Eagle Nebula. Squeezed in the grip of their own gravity, the globules grow still more dense. Their interiors heat as the density increases. Ultimately, they become sufficiently hot and dense for nuclear fusion, the mechanism that powers hydrogen bombs, to begin at their centers. Light and heat pour forth. A star is born.

Meanwhile, the density wave proceeds on its way, leaving new stars scattered behind it, like a planter broadcasting seed.

---

8 The Eagle Nebula (= M16 = NGC6611), a glowing cloud of gas, measures some seventy light-years in diameter; on the scale of this photograph our solar system would be a microscopic speck.

THE ROSETTE NEBULA

Newborn stars sputter and flash, pouring out light with a youthful profligacy that they cannot long maintain. The energy generated by new stars can have pronounced effects upon the remaining interstellar clouds surrounding them. The light and other energy generated by the more massive young stars may be sufficient to set up local shock waves that engender the collapse of segments of the surrounding cloud to form additional stars. If so, stars repay their debt to the interstellar medium from which they formed by assisting it in creating more stars.

In the Rosette Nebula we see a nest of young stars whose outpouring of energy appears to have swept away the remaining gas in their immediate vicinity, leaving the nebula hollow like an eggshell. The dark globules prominent in the surrounding cloud are probably congealing on their way to forming further stars. It may well be that the collapse of these portions of the cloud was triggered by the pressure of light from the furiously burning new stars at the center of the nebula.

---

9 The dark globules in the Rosette Nebula (= NGC2244) appear to be condensing to form stars.

34

10 A giant star-making machine, the Rosette Nebula wreathes some of the young stars that it created.

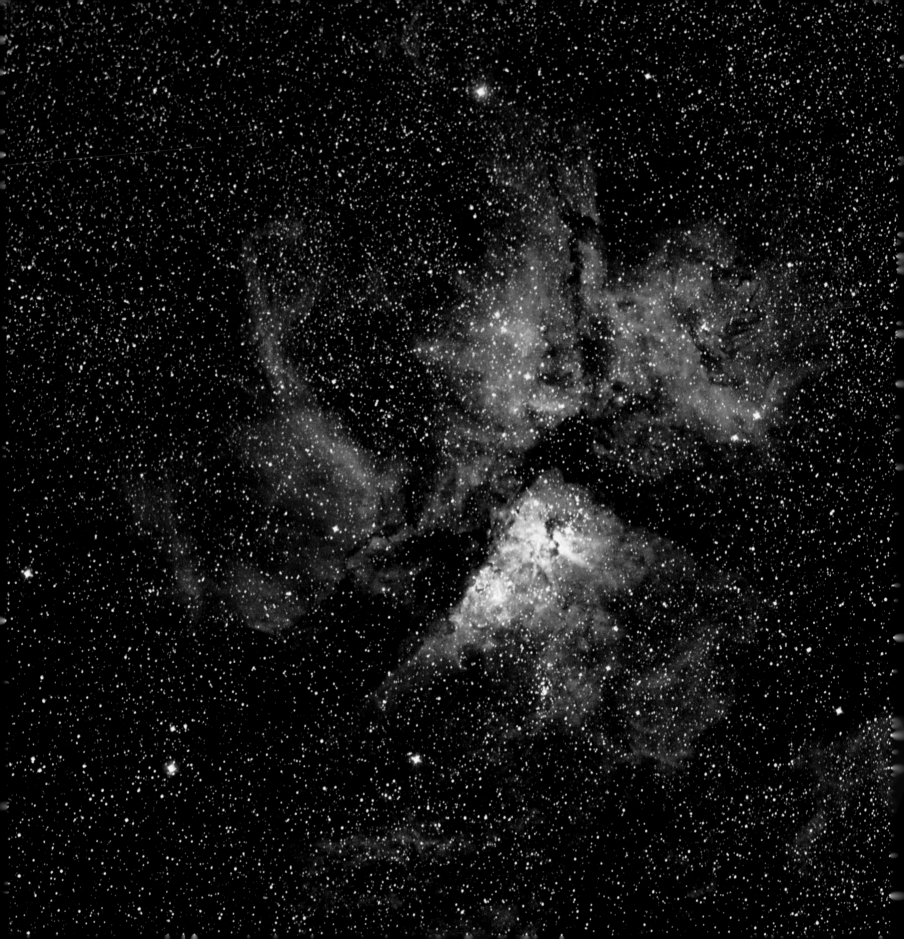

THE ETA CARINA NEBULA

As the sun pursues its orbit around the center of the Milky Way Galaxy, from time to time it passes through one of the galaxy's spiral arms. At the present epoch, when we humans have come upon the scene, built telescopes, and taken an interest in these matters, the sun and its planets lie near the inner edge of one of the arms. We have named it the Orion arm, after the bright constellation most of whose stars lie within it.

When we look in toward the center of our galaxy, we are confronted by the long flank of the next spiral arm in from the sun, which we have named the Sagittarius arm. Here again we see an array of bright nebulae, stretched out before us like the lights of a passing ocean liner. Among them are the Trifid, Lagoon and Eagle nebulae.

These glowing clouds are bright spots in an archipelago of dark gas and dust ranged along the length of the arm. The sinuous black veins winding through the star field in the photograph of the Eta Carina Nebula are not empty space, but are part of a meandering dark cloud to which the bright nebula belongs. The bright nebula stems from the dark cloud, like a blossom on a black branch.

---

11 The Eta Carina Nebula (= NGC3372), like most bright nebulae, is an illuminated portion of a larger dark cloud, seen here silhouetted against the background stars.

12 A shock wave may be seen moving like a storm front through the interstellar medium in this view of the interior of the Eta Carina Nebula.

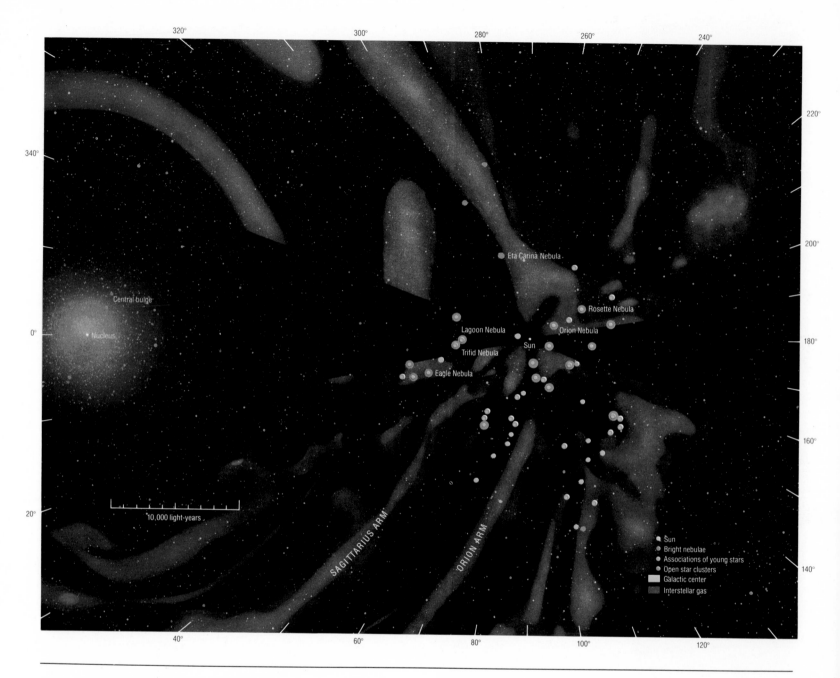

Figure 3. The Sun's Place Among the Spiral Arms of Our Galaxy
Beauty increases with proximity, says an informal maxim in astronomy, and as this map shows the beautiful bright nebulae whose photographs appear in this section belong to spiral arms of our galaxy that lie close to our solar system. The Lagoon, Trifid and Eagle nebulae are associated with the Sagittarius Arm, while the Eta Carina, Orion and Rosette nebulae are associated with the Orion Arm. The sun lies between these two arms, close to the inner edge of the Orion Arm. The cone-shaped zones to the left and right of the sun that interrupt the contours of the arms are not real, but indicate areas where intervening dark gas clouds block our view and prevent accurate mapping of the areas beyond. Like early maps of the New World, this map should be considered only a rough approximation; we are just beginning to learn our way around our own galaxy.

THE TRIFID AND LAGOON NEBULAE

These nebulae are three-dimensional structures. They have considerable depth as well as their more evident articulation in the two other dimensions. This is especially well illustrated by the Trifid Nebula. Starlight and glowing gas within light it up like a ship's lantern; the dark portions of the cloud that seem to divide it into thirds (hence the name "Trifid") are foreground parts of the cloud that we see in silhouette like struts in a lantern's globe. The red hues are produced by glowing hydrogen gas. The ice-blue regions are primarily dust particles in the cloud reflecting the light from stars in the nebula. These young stars, very hot, radiate generously in the energetic wave lengths of blue light.

The Lagoon Nebula was named by an observer who fancied that the dark rift running across its face resembled a map of a harbor. Actually this feature is almost certainly a dark foreground portion of the cloud, like the "struts" that cut across the Trifid Nebula.

Young stars abound in the Lagoon Nebula, many of them still flaring up erratically as they struggle to bring their gravitational and radiative forces into the balance that will be required for each to settle down into a tenure as a stable star. The intense light from these young stars and from the glowing rebula enshrouding them makes them some of the brightest objects in the galaxy.

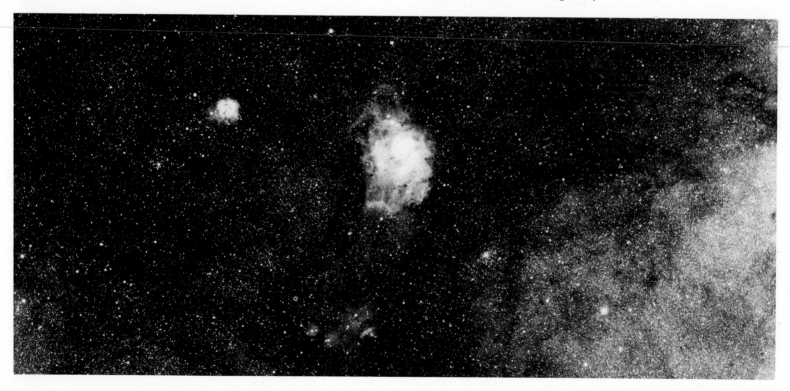

13 A field view of the Trifid and Lagoon nebulae suggests that both are tangled in the same large and otherwise dark interstellar cloud. A final determination of this awaits a better assessment of the distance separating the two nebulae, currently estimated to be as much as five thousand light-years or more. The Trifid, according to several estimates, lies considerably farther away from earth than does the Lagoon.

14 (overleaf left) The dark globules of the Lagoon Nebula (= M8 = NGC6523), hatcheries of young stars, are estimated at this stage in their collapse to measure a light-month or two in diameter.

15 (overleaf right) The Trifid Nebula (= M20 = NGC6514) belongs, like its neighbor bright nebula the Lagoon, to the Sagittarius arm of our galaxy.

OPEN STAR CLUSTERS

Stars that were born together stay together for a while, in associations known as star clusters. There are two distinct kinds of star clusters. Open clusters, like those seen on these pages, range in population from a few dozen to a few hundred stars. Globular clusters, like those seen on the following pages, are much larger and have populations that range into the millions of stars. Globular clusters are gravitationally fairly stable, almost like tiny galaxies all their own, and many are quite old. Open clusters, less permanent, typically are young. The stars of the Pleiades (page 49) are not much more than one hundred million years old.

The reason most open star clusters are young is that they cannot hold together for long. Lacking the imposing mutual gravitation of the hundreds of thousands of stars found in a globular cluster, they tend to fall apart as time goes by, losing their member stars by way of both internal and external influences. Stars on the outskirts of the cluster may be lost to the gravitational tug of a passing star, or of another star cluster, or of the galaxy as a whole. More often, stars are lost internally, when a lesser star in the cluster passes near one of its more massive compatriots; the gravity of the massive star accelerates the lesser one, like a skater in a crack-the-whip, hurling it out of the cluster.

The loss of each star reduces the gravitational potential of the cluster as a whole, leaving it increasingly vulnerable to further defections. Ultimately the cluster is reduced to only a skeleton crew and may dissipate entirely into the general galactic population. Most of the stars we see in the galaxy today, even solitary stars like our sun, may once have belonged to open clusters.

---

16 The young cluster NGC3293 lies about ten thousand light-years from earth.

17 The light of stars in the Pleiades Cluster (= M45 = NGC1435), reflecting off the dust cloud in which the cluster is embedded, forms a veil that glitters like diamonds.

The more than one hundred globular star clusters associated with our galaxy have two particularly intriguing characteristics. One is their age. Globular clusters typically are very old. Some have been estimated to be over fifteen billion years old, a longevity comparable to that of the galaxy itself. The other noteworthy characteristic of the globular clusters is their distribution in space. While most of the bright stars of our galaxy are situated along the plane of its disk, globular clusters are found above and below the plane as well.

These two attributes of globular clusters are of great interest to researchers concerned with how the galaxy may have formed. Most of these theorists have determined, by reconstructing the physics of the situation, that our galaxy began as a more or less spherical aggregation of gas that subsequently collapsed to form the disk it describes today. Opinions differ as to how long such a collapse took to occur, and as to how many stars had formed from the primordial gas by the time of collapse, but most agree that long ago the Milky Way protogalaxy was roughly spherical in shape. This impression gains credence from the fact that the globular clusters, made up of very old stars, are still found well out of the galactic plane, distributed in a spherical volume of space that may replicate the dimensions of the old Milky Way protogalaxy. The distribution of globular clusters may constitute a relic of the way the galaxy once was, like the charred frame of a building whose walls have collapsed in a fire.

The clusters themselves are imposing. The largest globulars have millions of stars and, at least in the case of those that lie well away from their parent galaxy, it is difficult to decide just where to draw the line distinguishing a large globular cluster from a dwarf galaxy.

Dramatic perspectives on the cosmos are implied by the location of many globular clusters and by their wealth of stars. Imagine, for instance, that the sun and earth were situated not here in the plane of our galaxy, but in the outer fringes of a remote globular cluster that lies well away from the plane. For a half of each year our night skies would be jammed with the brilliant stars of our home cluster, their light so intense that darkness would never really fall. For the other half of the year, during the seasons that found us on the side of the sun away from the globular cluster and toward the Milky Way Galaxy, we would see the galaxy laid out flat from one horizon to the other.

A price paid by remote globular clusters for their enviable

18 The light from old red giant stars warms the hues of the globular cluster NGC2808, twenty-five thousand light-years away.

19 The globular cluster M13 (= NGC6205) is approximately two hundred light-years in diameter, but most of its more than one million stars occupy a central region whose diameter is under one hundred light-years; in these relatively crowded quarters the average clearance is one star per cubic light-year.

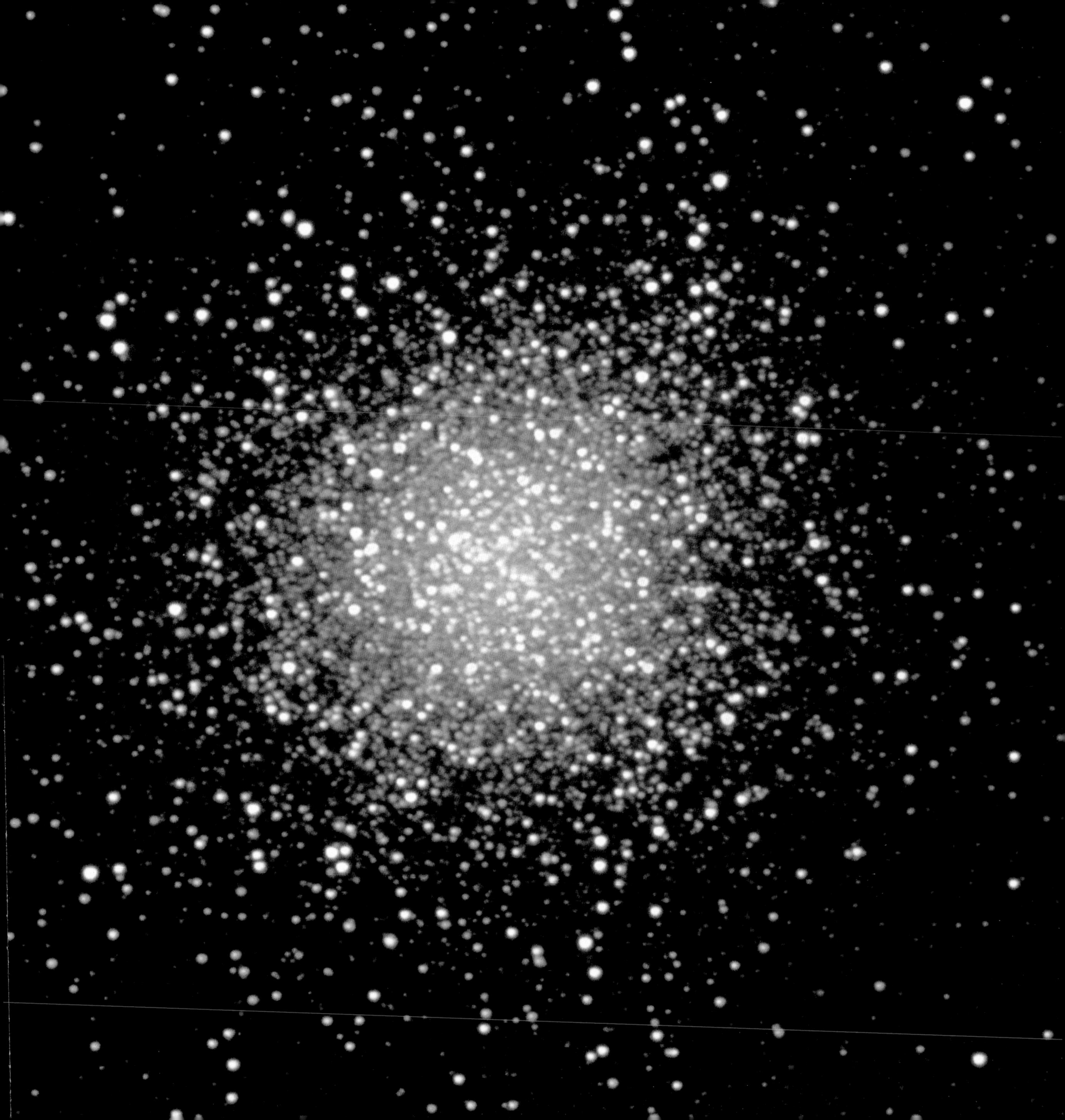

view of the cosmos is sterility. The interstellar gas and dust needed to make new stars is concentrated along the galactic plane. Globular clusters lying far from the plane are denied these riches, and so stars rarely form there. The stars we see in globular clusters are mostly old survivors born long ago. The virtues of these gerontocracies are those of the baroque; they are refined and enduring, but in terms of stellar evolution they represent nearly the end of the line.

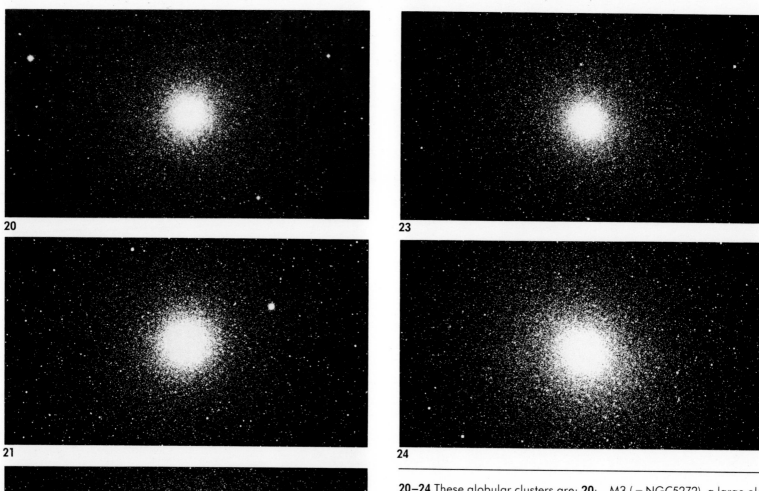

20

21

22

23

24

20–24 These globular clusters are: **20:** M3 (=NGC5272), a large old cluster with an estimated five hundred thousand stars, many of them dwarves and other degenerates that have seen better days; **21:** M15 (=NGC7078), almost fifty thousand light-years distant and well away from the plane of our galaxy; **22:** Omega Centauri (=NGC5139), the brightest known globular cluster, which lies near the plane of the galaxy halfway in from the sun toward its center; **23, 24:** M5 (=NGC5904) and 47 Tucana (=NGC104), which both show evidence of flattening perhaps due to rotation.

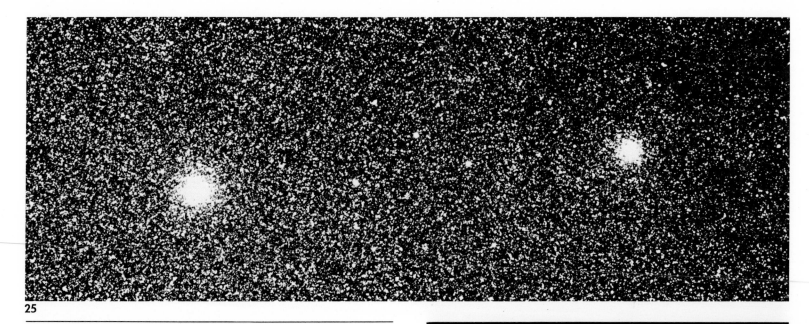

25

25, 26 The globular clusters NGC6522 and NGC6528 (above) lie within the plane of our galaxy, passing their time in the company of billions of noncluster stars. The remote globular cluster NGC2419 (right), on the other hand, has wandered more than three hundred thousand light-years from the center of our galaxy, to a point near the limits of the galaxy's gravitational domain; this "intergalactic tramp" cluster follows a lazy orbit around the galaxy that will require over three billion years to complete.

26

The Death of Stars

The lesson of life that nothing lasts forever is echoed in the death of stars. Though symbols of constancy, stars ultimately must perish. And, as we find elsewhere in nature, the mortality of the individual plays a part in the evolution of the whole.

Stars die in accordance with how they lived. Modest stars like our sun conclude their careers modestly. Having depleted most of their fuel, they expand to become blowsy, dimly glowing giants that shrug off their outer atmospheres and settle down into retirement as dwarf stars. More massive stars conclude their brilliant reign in a more spectacular fashion—they explode. Extraordinarily massive stars wind up their extravagant careers by exploding with extraordinary force.

This suitability of death to life in stars pertains also in the design of stellar tombs. The remains of ordinary stars take the form of inconspicuous dwarves. More massive stars collapse to form more noticeable monuments, the neutron stars, whirling stellar cinders compressed to a state harder than diamond. The most imposing stars may collapse so forcefully as to form black holes, cutting themselves off from the rest of the universe and realizing the ambitions of the Pharaohs in that their remains can never be exhumed.

process may be described in a radically simplified fashion by saying that the core, its fuel spent, cools and collapses into a dwarf remnant while the massive outer envelope of the star—the part that had surrounded the core—boils away into space under the impulse of its own heat.

The so-called "planetary" nebulae like these are stars that we have caught in the act of discharging such a shell of gas. What appears to be a ring surrounding the star is actually a thick shell or bubble. As the shell expands into space, light from the degenerate star it left behind excites the gas to glow. The delicate colors of the shell result from the excitation of its various gases, among them hydrogen, oxygen and nitrogen. The term "planetary" is another of those misnomers that complicate astronomical lexicography; it results from an error by early astronomers, who, equipped with only small telescopes, noticed a vague resemblance between these shells of gas and the disks of the planets of our solar system.

"Planetary" nebulae are passing things. Each will go on expanding until it has dissipated into interstellar space. The "planetary" nebulae seen in these photographs are typically only a few tens of thousands of years old; in another few tens of thousands of years they will have all but vanished. Meanwhile, the core of the star left behind will in most cases have settled down as a dwarf, able to glow dimly for a very long time before itself fading away. The "planetary" nebula phase therefore represents a brief episode late in the life of a star, one that costs it no greater portion of its time than a nonfatal heart attack deducts from the life of a human being.

Each year in the Milky Way, a few aging stars slough off their skin in this manner. The result is that thousands of "planetary" nebulae ornament our galaxy at any given time. In all, they add enough material to the interstellar medium each year to form five new stars the mass of the sun.

ERUPTING AND EXPLODING STARS

Novae and supernovae are stars that divest themselves of material not with the orderly demeanor of a "planetary" nebula but with explosive violence. The name "nova," meaning new star, is an index to the spectacle they create; exploding as a nova or supernova, a previously nondescript star can flare up so brightly that it dominates the sky, creating the illusion that a new star has come into being where none was before.

The terms "novae" and "supernovae," however, cover a considerable range of stellar violence; a very minor nova may not be much more traumatic to the star involved than is the ejection of gas to form a "planetary" nebula, while supernovae may attain apocalyptic levels of violence. When the mechanisms of stellar explosions are better understood, it may be possible to assign supernovae, novae and the events that produce "planetary" nebulae places within a single spectrum describing the means by which stars divest themselves of excess baggage on the way to their graves.

Most spectacular of these stellar death rites are the supernovae. Their luminosity may surpass that of billions of normal stars, and they can eject enough material to make several of our solar systems. Such an event would spell disaster (literally, for the word "disaster" comes from the Latin for "unfavorable aspect of a star") for anyone having the misfortune to live in the celestial neighborhood of a supernova. Planets of the exploding star would be vaporized, and planets of other stars lying within a few light-years would be bathed in a quantity of radiation sufficient to sterilize them. Yet, as we find with terrestrial disasters like forest fires, earthquakes and hurricanes, the destructive forces of nature contain seeds of creation. Exploding stars play a vital role in the ecology of the galaxy.

Supernovae build heavy elements. All stars are engaged to some degree in evolving complicated atoms out of the simple atoms of hydrogen and helium. Generally speaking, the more massive a star and the hotter its interior, the heavier the elements it can construct there. Very massive stars are capable of forging atoms as heavy as those of iron. But they are unable to go further. Iron atoms, highly stable, cannot be broken down and rebuilt into heavier atoms even in the interior of a hot star. To surmount this barrier calls for a drastic solution, and a supernova is just that. In its intense heat a wide array of heavy atoms are forged, then sprayed into interstellar space.

So supernovae are not solely celestial death spasms, but represent the crowning achievement of a star that caps a lifetime of element-building by seeding its cosmic neighborhood with atoms as exotic as those of uranium and gold. Stars and planets that form subsequent to the supernova inherit these heavy elements as part of the interstellar cloud from which they are made. That is how we inherited the heavy elements we find here on earth: They had been injected into the interstellar medium by stars that exploded before the sun and earth condensed from that medium.

It is curious to reflect that the materials employed to build the telescopes that took the photographs on these pages, and the carbon used to ink these words, were created within

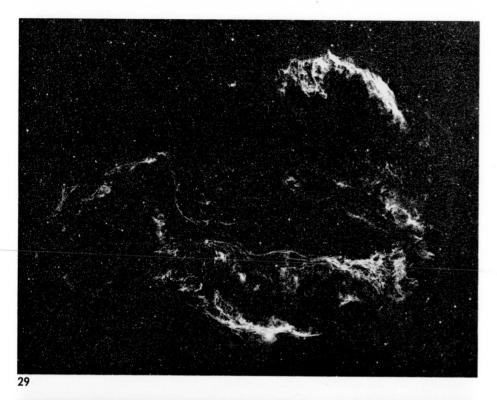

29 The Veil Nebula (=NGC6960/6992), ejecta from a star that exploded thirty or forty thousand years ago, is now nearly one hundred light-years in diameter. It continues to swell, a slowly bursting bubble, its rate of expansion decreasing as it entangles itself with other interstellar clouds. Rich in heavy elements, it seeds the interstellar medium like a gigantic thistledown.

29

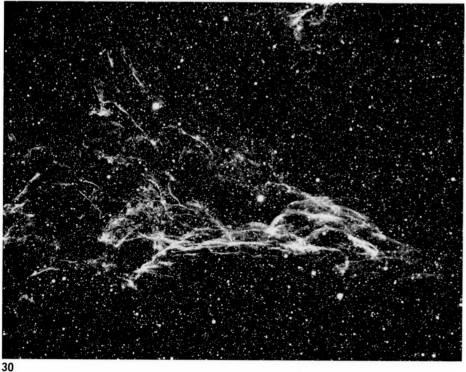

30 A few of the streamers of the aptly named Veil Nebula are seen here in detail; the segment visible in this photograph measures approximately twenty light-years across.

30

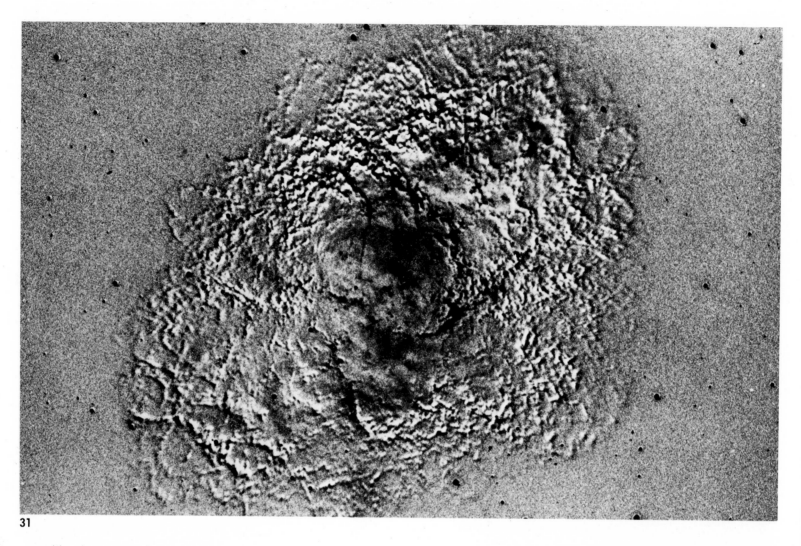

31

stars like those we now contemplate. The stars facilitate our study of them.

When a star explodes it leaves behind two sorts of debris —that which it has blown into space and that which stays behind. The material blasted into space expands until it has dissipated and mingled with the rest of the interstellar medium. The material left behind collapses to form a dwarf star, or a denser object known as a neutron star, or an object so dense that it swallows up its own light and becomes a black hole.

The Crab Nebula (right) is the remnant of a supernova that occurred only five thousand light-years from the sun, quite recently in cosmic time; its light reached earth in July 1054 and was noted by Chinese, Arab and American Indian observers. Thanks to its proximity, the Crab Nebula can be

studied in detail. It provides us, so to speak, with a warm body on which to conduct a supernova postmortem.

The crushed cinder at its center, all that remains of the star,

---

31 Expanding at a rate of six hundred miles per second, the Crab Nebula supernova remnant can be seen to grow in size as the years go by. Here a positive print, in which the nebula appears light in color, has been superimposed on a negative print made fourteen years later, in 1964. Expansion of the cloud during those fourteen years has offset the dark from the light images, producing a bas-relief effect.

32 Wreckage of an exploded star, the Crab Nebula (=M1=NGC1952) consists of an envelope of material blasted into space when the star blew up, and at its center a collapsed dwarf, or neutron, star.

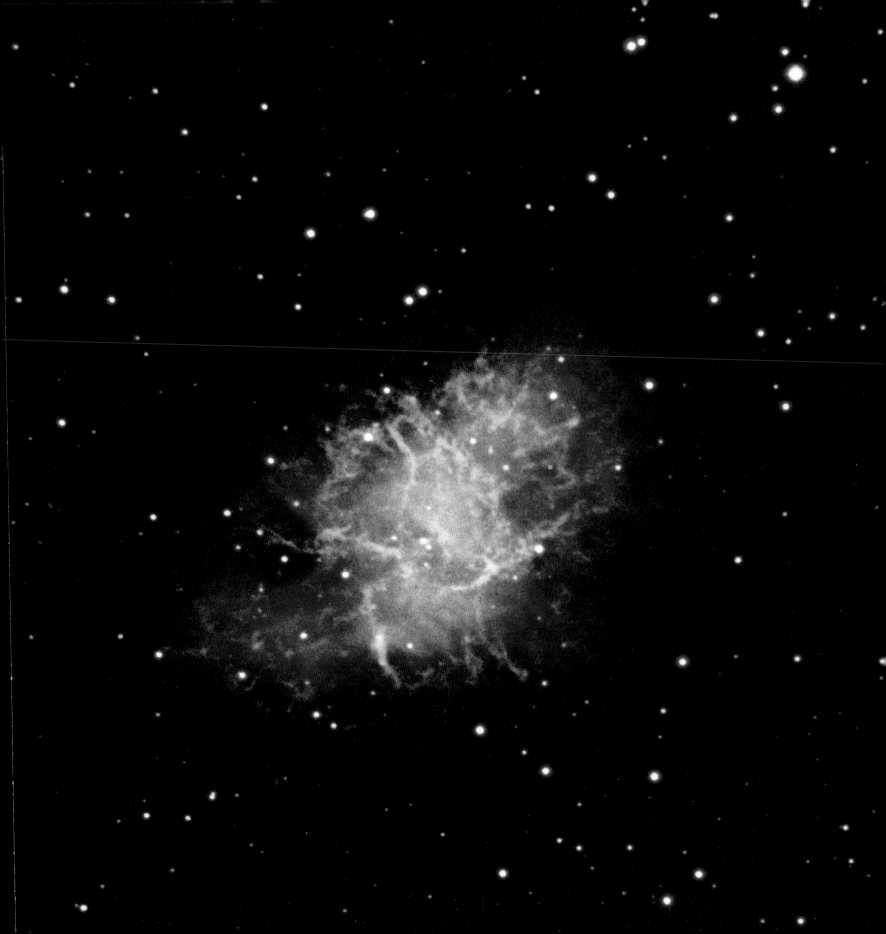

is compressed to such a density that a spoonful of it would weigh millions of tons. Spinning thirty times per second, this strange object emits pulses of energy in both radio and optical wavelengths; astronomers call it a neutron star, since it consists primarily of neutrons packed closely together, or a pulsar, when referring to the pulses of radio waves it emits.

The surrounding envelope of material hurled off in the explosion glows with excitation produced by radiation pouring out from the cinder. The colors of its filaments come principally from ionized hydrogen, carbon and sulphur.

The effects of a supernova like the one that created the Crab Nebula are varied and subtle. Some are intimate; high-energy particles of the sort we call cosmic rays are produced by the remnant and some of these strike the earth, where they are capable of shattering genetic material in reproductive cells and creating mutations that alter, however slightly, the course of the evolution of life here. No less important in the history of humankind are the intellectual effects produced by the sight of a supernova, as when the Danish astronomer Tycho Brahe saw a supernova in 1572 as a young man in his twenties and was emboldened to question the philosophical authority of Aristotle, who had held the realm of the stars to be eternally unchanging.

BLACK HOLES

The cosmos is so abundant in things we can see—stars, bright nebulae, the planets here in our solar system—that we may neglect to consider how much there is in the cosmos that we do not see. A considerable portion of the mass of the universe, perhaps most of it, is inconspicuous or even invisible.

Most of the inconspicuous constituents of the cosmos ought to be detectable sooner or later, given good enough equipment. Planets of other stars, undetectable at present, ought to be discernible with specially designed telescopes operating in space. Dwarf stars too dim to be photographed with existing telescopes can be expected to evidence themselves when telescopes with far greater light-gathering power have been built. Matter in highly rarified forms, such as the ethereal clouds of cold hydrogen gas found in intergalactic space, can be located by charting energy emitted by their scattered atoms at the invisible wavelengths of radio or X-rays.

But there is one form that matter may take that renders it genuinely invisible, hides it away from view forever—a black hole.

The term "black hole" describes the outward appearance of a class of astronomical objects that are compressed to such a high density that their gravitational field prevents even their own light from escaping them. A black hole has in effect retired from the outside world. No energy of any sort emerges from it. A black hole offers us no picture of itself, and if we tried to take a flash photograph of it, it would simply swallow the light from the flash. Hence the name.

Theoretical astrophysicists envision several ways that black holes might come into being. One is by way of the collapse of the core of a star. We have seen that massive stars that have exploded as supernovae leave behind degenerate cores that, caught in the grip of their own gravitation and no longer propped up by the energy generated by nuclear fusion, can collapse to form the highly condensed objects known as neutron stars. But if the core left behind after a supernova is sufficiently massive (more than about three times the mass of the sun), it could collapse right through the neutron star stage to become so small and gravitationally powerful that it swallows up the light it produces.

Another possibility is that black holes might be formed by the collapse of still more massive objects, such as the nuclei of galaxies or of large globular star clusters. Or one can conceive of mini-black holes, tiny as subatomic particles.

These are the theoretical possibilities. Whether black holes exist, and in what abundance, depends upon the extent to which nature has pursued the paths theoretically available to her. Our experience here on earth indicates that nature does pursue many exotic paths (e.g., bower birds), while declining to pursue others (e.g., unicorns), and by extrapolation beyond the earth we can imagine an enormously heightened variety (e.g., extraterrestrial unicorns?).

If massive stars do collapse into black holes, then we can estimate that there are some one hundred million black holes in the Milky Way today, each with a mass of at least several times that of our sun, each the remnant of a huge star that exploded in the past. If, in addition, the nuclei of many galaxies harbor black holes, we would expect each to weigh tens of thousands of times the mass of the sun, owing to the abundant opportunities for these black holes to gobble up interstellar gas and other material available to them in the crowded central galactic regions. And if, to take the extreme case, those scientists who envision ubiquitous subatomic black holes are correct, more than ninety percent of the mass of the universe might be cached away in the realm of the invisible, demoting the visible cosmos to the status of something of an afterthought. It remains for future observation to determine where the truth lies along the spectrum ranging from a universe with no black holes in it to a universe composed mostly of black holes.

To search for a black hole is to embark upon a quest of the sort that would have delighted Lewis Carroll. One is looking for the invisible. How to go about it?

A promising strategy is to seek evidence of the effect of black holes upon their surroundings. This might be called the Invisible Man Approach, after the character in the H. G. Wells science fiction novel who is safe so long as he does nothing to interact with his environment, but risks detection if he blunders into someone or tries to make off with a sandwich. Black holes, infinitely hungry, will consume any material that comes near enough for them to trap it. A black hole embedded in an interstellar cloud will swallow portions of that cloud. A black hole in a double star system will, if it gets close enough to its companion star, strip material from the companion and consume it. A black hole feeding ground ought to be detectable by observing energy, notably X-rays, thrown off by the doomed material as it swirls down into the black hole.

Orbiting X-ray telescopes have succeeded in locating several X-ray sources that very possibly represent black holes. The first such candidate to be discovered was Cygnus X-1. Here the black hole, if it is such, is part of a double star system something over six thousand light-years away. The visible star in the system has a mass of thirty times the sun's, while its companion, alleged to be a black hole, has about six times the sun's mass.

Other powerful X-ray sources have been located at sites where we might expect to find black holes. These include the nuclei of some galaxies and of massive globular clusters.

The "event horizon" of a black hole, the boundary from within which nothing can escape, appears to be inviolable. This may be why black holes are so provocative to the human imagination, for few if any absolute boundaries have yet been found in nature. Human history is dotted with stories of the surmounting of them—the edge of the world, the speed of sound, travel in space—and an immediate human response to being told that no one can cross the event horizon of a black hole and return is to try to find a way to beat the system.

Structure of the Milky Way

As we survey the stars and interstellar clouds surrounding us, we see that they are organized not randomly but in accordance with a pattern. Most are incorporated into a flattened disk that we now know to be the disk of our galaxy. Its appearance from our vantage point is that of a broad river of light stretching across the sky and glowing with the combined light of myriad stars—the Milky Way. Once we learned that it was in fact our view of our own galaxy, we named the galaxy after it.

The photographs on the following pages show portions of the Milky Way that lie toward the center of our galaxy. Our sun is more than halfway out from the center, so the richest star fields from our vantage point are those we see when we look back toward the center rather than out away from it.

Our appreciation of the view is enhanced if we strive to imagine it in three dimensions. The interstellar clouds display clear evidence of dimensionality; their loops and festoons can be seen to have depth, like the curving ribs of a skeleton. The concentration of interstellar material along the plane of the galaxy is typical of spiral galaxies; compare this view with that of the external galaxy (NGC2683) seen edge-on (page 111).

---

33 (overleaf left) Rich fields of stars and interstellar gas and dust characterize the Milky Way in the direction of the constellation Cygnus. Here is to be found an enormous gas bubble apparently produced by a series of exploding stars, and the X-ray source Cygnus X-1, thought to be a black hole.

34 (overleaf right) The Milky Way. When we look toward the center of our galaxy, we see hundreds of millions of stars that, like our sun, lie along the galaxy's flattened plane. These portions of the Milky Way are found in the southern skies of earth in the direction of the constellation Sagittarius.

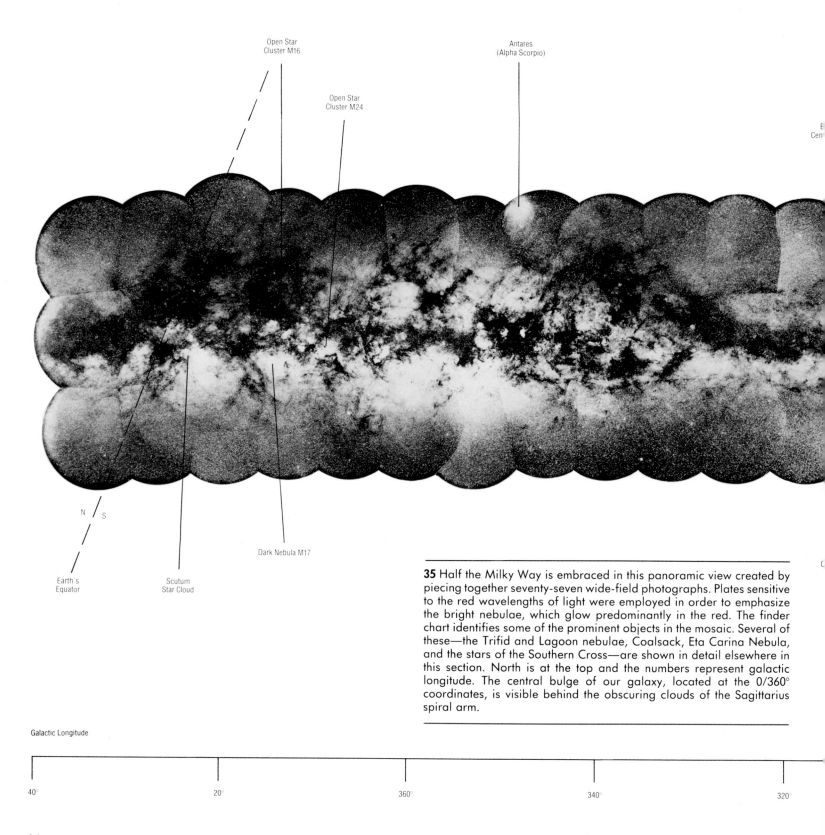

Open Star
Cluster M16

Open Star
Cluster M24

Antares
(Alpha Scorpio)

B
Cen

Dark Nebula M17

N / S

Earth's
Equator

Scutum
Star Cloud

35 Half the Milky Way is embraced in this panoramic view created by piecing together seventy-seven wide-field photographs. Plates sensitive to the red wavelengths of light were employed in order to emphasize the bright nebulae, which glow predominantly in the red. The finder chart identifies some of the prominent objects in the mosaic. Several of these—the Trifid and Lagoon nebulae, Coalsack, Eta Carina Nebula, and the stars of the Southern Cross—are shown in detail elsewhere in this section. North is at the top and the numbers represent galactic longitude. The central bulge of our galaxy, located at the 0/360° coordinates, is visible behind the obscuring clouds of the Sagittarius spiral arm.

Galactic Longitude

40°   20°   360°   340°   320°

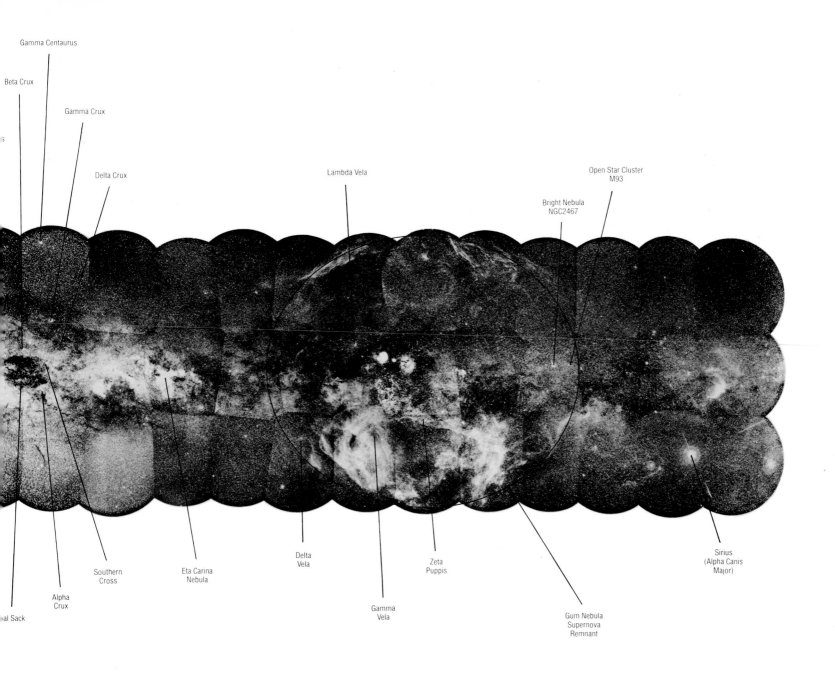

Gamma Centaurus

Beta Crux

Gamma Crux

Delta Crux

Lambda Vela

Open Star Cluster
M93

Bright Nebula
NGC2467

Southern
Cross

Eta Carina
Nebula

Delta
Vela

Zeta
Puppis

Sirius
(Alpha Canis
Major)

Alpha
Crux

al Sack

Gamma
Vela

Gum Nebula
Supernova
Remnant

300°   280°   260°   240°   220°

65

II/The Local Group of Galaxies

He showed me a little thing,
the quantity of an hazel-nut,
in the palm of my hand;
and it was as round as a ball.
I looked thereupon with the eye
of my understanding, and thought:
What may this be? And it was
answered generally thus: It is
all that is made.

—ST. JULIANA

A Journey Out of Our Galaxy

We speed out of the Milky Way in our imaginary spaceship like divers ascending from the depth of the sea. The myriad bright stars which had been our companions now diminish in number, then fall away behind us. In their stead we are left with the scattered stars of the galactic halo. Most are dim dwarves, remnants of stars that formed more than ten billion years ago, when the infant galaxy was more nearly spherical and had not yet collapsed to its present flattened shape. A few "runaways," younger stars ejected out of the plane of the galaxy by quirk gravitational encounters, flit among these elders like bright tropical fish venturing from their accustomed shallows.

To exit a galaxy is no handy matter, but eventually we attain a sufficient remove to be able to view the galaxy spread out below us. The central bulge of the galaxy looms directly under us, its shape and color like that of a hill of sand. The galactic disk surrounds it, a monumental tangle stretching to the celestial horizons. The glowing clouds of the spiral arms wend their way out through the disk, often obscured by intervening dark clouds, like a river cutting through a jungle. Here and there dark tattered towers are reared up out of the welter of the disk, masses of interstellar gas and dust that have been heaved out of the galactic plane in the course of the collisions of clouds and the explosions of stars.

We climb further and our view of the galactic disk improves. The time comes when we can discern the sun, a little yellow star nestled in the embrace of the outer reaches of one of the spiral arms, a dot of light barely visible through the ship's telescope. Here long ago was our home.

The globular star clusters make for spectacles close at hand. From time to time one of these chandeliers of stars passes abeam of our ship. We are tempted to stop and explore its hundreds of thousands of stars, but we are bound for territories more remote.

When the outermost of the globular clusters along our course has fallen away aft, we celebrate having left our home galaxy behind. The choice of demarcation is rather arbitrary, for we still lie well within the gravitational domain of our galaxy. But we feel the need of a cheering toast, for we are embarking upon the awful gulf of some of the emptiest space known in a universe that is mostly space. Our galaxy hangs behind us like a gong, its slowly diminished starlight painting shadows across our ship from aft, while ahead yawns the void, its only light the pearlescent background haze of a universe of galaxies.

Our eyes seek out landmarks lest we be seized by vertigo. Well off to port hangs the most evident galaxy in sight after the Milky Way, the Large Magellanic Cloud. Beyond it we can make out the less orderly patches of starlight that comprise the Small Magellanic Cloud and the Sculptor and Fornax dwarf galaxies. To starboard lie two other dwarves, the little Leo I and Leo II galaxies. We steer for them.

Seven hundred fifty thousand light-years separate the Milky Way from the Leo pair. Our activities during this phase of the voyage are those suitable to a long haul. We carve scrimshaw, repair our gear, read all the back issues of the National Geographic. Down below the stokers shovel whole planets' worth of fuel into the engines to maintain our acceleration. Leo I and II slowly grow in the forward viewports.

Now we can look back upon most of the Local Group in a single gaze.

The Milky Way Galaxy, though still imposing, has shrunk until it covers less than ten degrees of the sky; we can eclipse it with an outstretched hand. A little train of satellite galaxies stretches off to one side of the Milky Way. In the same part of the sky but far deeper in space hangs the spiral galaxy M33, and near it the majestic M31, dominant galaxy of the Local Group. Beyond them we can glimpse the elliptical galaxy Maffei I and its spiral companion Maffei II.

Will we in our little ship feel a last pang of leave-taking as we say goodbye to this corner of the universe, with its trillions of suns, in the light of one of which we came into being? Or has this already become too strange and remote to retain any of the warmth of home? We rocket past the Leo I dwarf galaxy and head out of the Local Group.

II/THE LOCAL GROUP

The Magellanic Clouds

The nearest galaxies to ours are the Magellanic Clouds. They are called "Magellanic" by virtue of their having been introduced to Western civilization by the crew of Ferdinand Magellan, whose circumnavigation of the earth took his ships beneath southern skies, where the clouds are to be seen. They are called "clouds" owing to their soft outlines and glowing appearance, which make them look something like scraps detached from the Milky Way. In the early decades of the twentieth century, a number of variable stars of the sort known as Cepheid variables, invaluable to astronomers as distance indicators, were identified in the Magellanic Clouds. Their discovery made it possible to establish that the clouds were too distant to be part of our galaxy, and had to be galaxies in their own right. The Large Magellanic Cloud is about one hundred fifty thousand light-years, the Small Magellanic Cloud about 250,000 light-years from the sun. Less than one hundred thousand light-years separates the two clouds.

The clouds lie well within the gravitational field of our galaxy and orbit it as satellites. This arrangement, small galaxies playing court to large ones, is commonplace in the universe; a major spiral like the Milky Way typically plays host to several satellites. In the case of the Milky Way, two of these satellites—the Magellanic Clouds—are considerably larger than the others. The large cloud has about fifteen billion stars, the small cloud about five billion.

The orbits of the Magellanic Clouds are marked by an enormous river of cold hydrogen gas, the Magellanic Stream, in which they swim. At least two other satellites of our galaxy —the Draco and Ursa Major dwarf galaxies—also follow orbits that lie within this river of gas. It is composed of at least six clouds connected by wisps of thinner gas. Each has enough gas to make several tens of millions of stars the mass of the sun. Doubt remains as to the origin of the stream. It may be composed of gas that was drawn from the Milky Way Galaxy by gravitational interactions with the Magellanic Clouds. Or it may be composed of primordial gas that, in a gravitational tug of war between the clouds and the Milky Way, has been permitted to settle into allegiance with neither. Similar cold clouds have been found in the spaces surrounding other galaxies, often in association with their satellites.

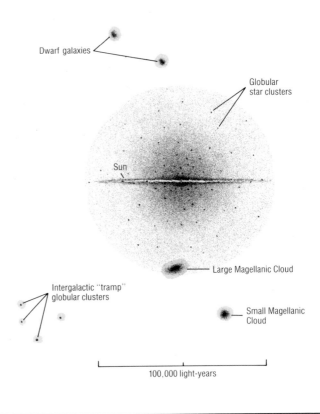

Figure 4. Location of Magellanic Clouds
In the outer domain of the Milky Way lie dwarf satellite galaxies and a few distant globular star clusters. The two large satellites known as the Magellanic Clouds lie in the foreground in this perspective, and should be envisioned as hovering a few inches above the page.

36 The large Magellanic Cloud, visible in great detail at its distance of only one hundred and fifty thousand light-years, shows evidence of many features familiar to us in our own galaxy; the ruddy cloud of gas toward one end of the galaxy, the Tarantula Nebula, belongs to the same species as the Orion and Trifid nebulae here in the Milky Way; with a diameter of eight hundred light-years and a mass of perhaps three hundred thousand times that of the sun, it is the largest such nebula known. Were it as close to us as Orion its brightness would rival the moon's.

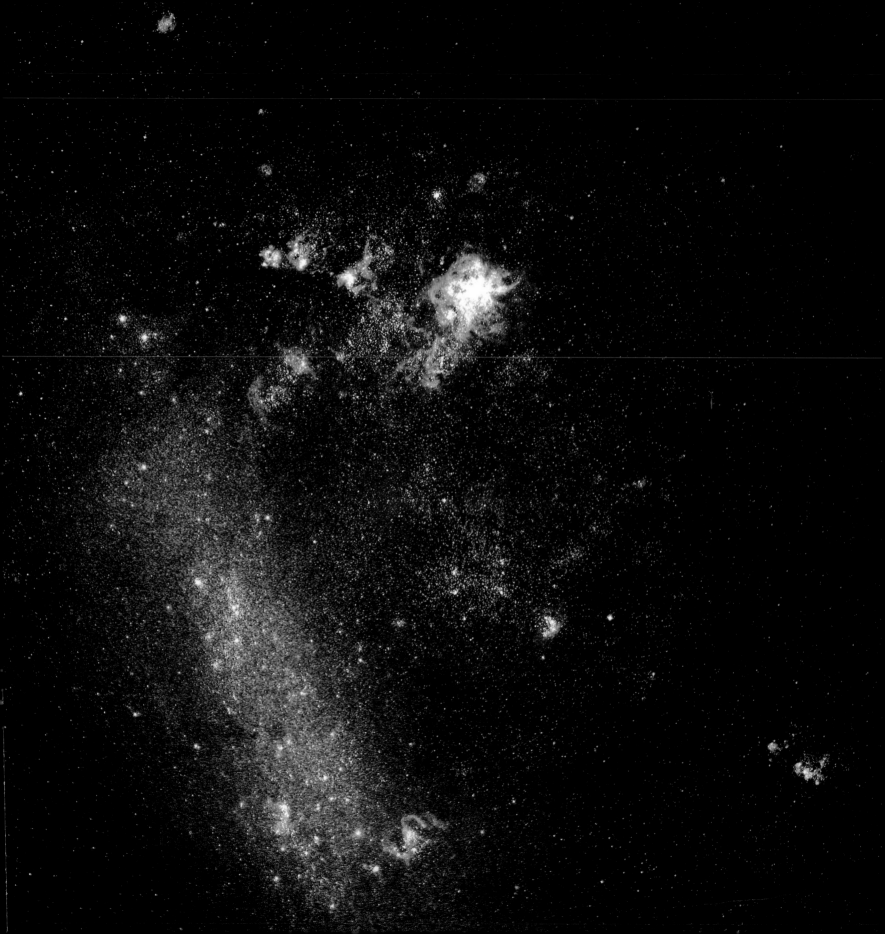

Figure 5

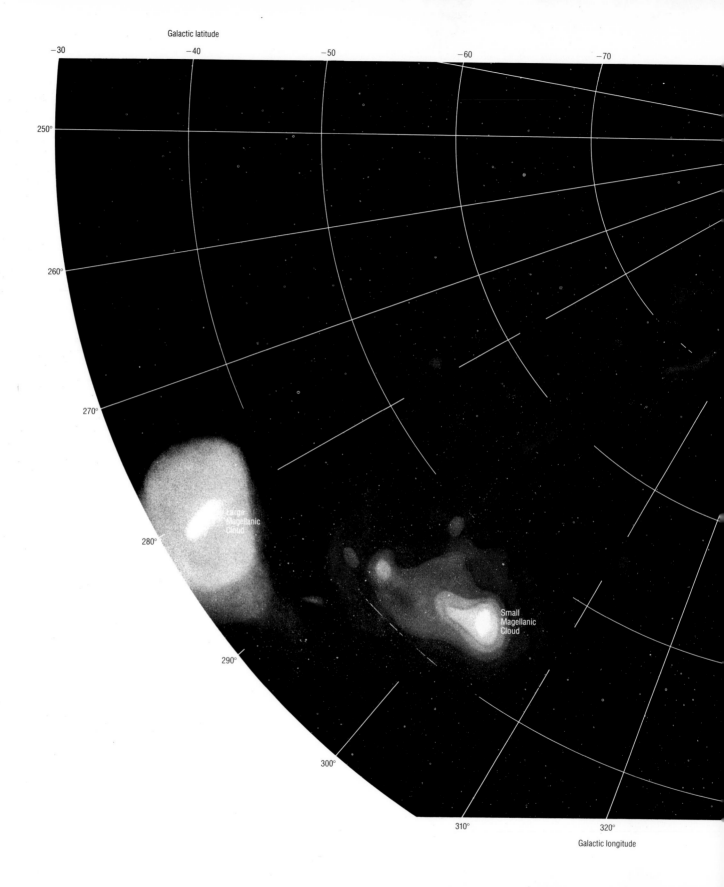

Galactic latitude

−30   −40   −50   −60   −70

250°

260°

270°

280°

Large
Magellanic
Cloud

290°

Small
Magellanic
Cloud

300°

310°   320°

Galactic longitude

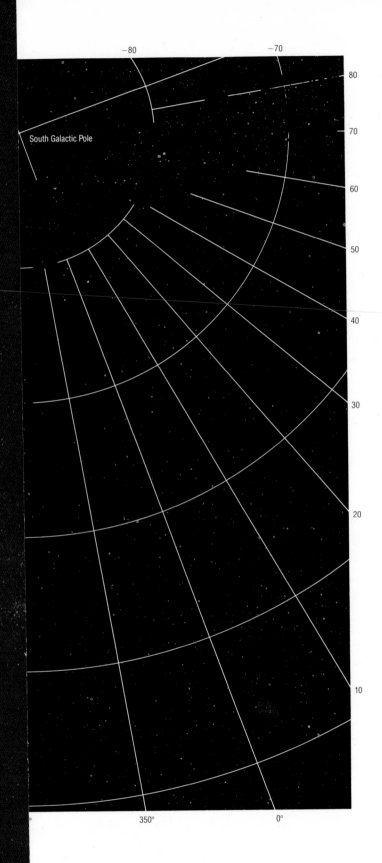

South Galactic Pole

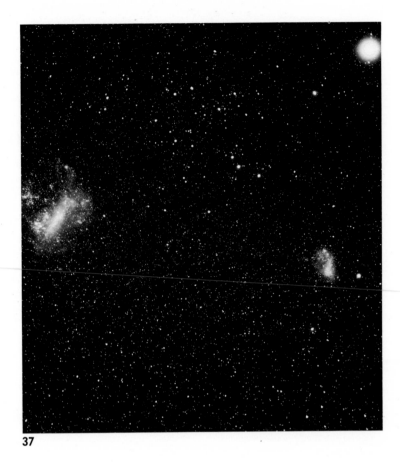

37

Figure 5. The Magellanic Stream
A giant loop of cold hydrogen gas encircling our galaxy, the Magellanic Stream, was so named because it lies along the orbit described by the Magellanic Clouds. The clouds appear to the lower left of this map, which was made by charting radio radiation emitted by the intergalactic gas. The coordinates are those of galactic latitude and longitude.

37 Riding at close quarters to each other, the Large and Small Magellanic Clouds are perhaps destined to merge into a single galaxy.

The Andromeda Galaxy

The Milky Way and its sister the Andromeda Galaxy constitute an example of one of nature's most grandiose creations, a pair of spiral galaxies. Many spirals belong to such pairs. Usually, the pairs are asymmetrical, like a crab's claws, one galaxy larger than the other. Andromeda, the larger of this pair, has about twice the mass of the Milky Way. The two galaxies rotate in complementary directions, one clockwise and the other counterclockwise, so to speak; this characteristic of their relationship, found in many other pairs of spirals as well, lends support to the hypothesis that the two galaxies formed at about the same time, from two adjacent whirlpools of primordial gas, rather than having formed far apart and later blundering into each other's company.

Similarities between the two galaxies are abundant. Each displays the accoutrements of a major spiral—a central region composed chiefly of old stars, an expansive flat disk populated by tens of billions of stars of widely assorted ages and chemical compositions, dust-laden spiral arms rendered incandescent by the light of newly formed stars, and a spherical halo of old dwarf stars embracing the galaxy and highlighted by hundreds of globular clusters. Each galaxy is attended by two prominent satellite galaxies plus many less prominent ones. It even happens that the plane of each galaxy is inclined to the other's line of sight at almost exactly the same angle, so that the Milky Way viewed from Andromeda ought to look quite a bit like Andromeda viewed from the Milky Way.

Here, while we behold a major galaxy at close range, may be an appropriate time to ponder what it is we are seeing in photographs of galaxies. The photograph records light and only light. Here on earth we are accustomed to seeing things by virtue of their reflected light; that is how we discern a smile or a mango or the moon. Photographs of galaxies offer us little such reflected light. Planets of stars in other galaxies, however many there may be, are too small and inconspicuous for us to see them, and light reflected from dust clouds—a phenomenon found occasionally in our galaxy, as in the case of the Pleiades star cluster—is too dim to show up in photographs like these.

What we do see is the light from billions of giant stars and from the bright nebulae in which some of the stars are embedded. This light reaches us from widely scattered quarters of what is, after all, a staggeringly large system of stars. If we consider that the Andromeda spiral is roughly one hundred thousand light-years in diameter, and note that its inclination to our line of sight sets the near edge of the spiral nearly one hundred thousand light-years closer to us than the far edge, then we face the curious conclusion that the light recorded in the photograph from the far edge is one hundred thousand years older than the light from the near edge. We are looking not only at a great deal of space, but at a thousand centuries of time.

The astronomer and philosopher of science Sir Arthur Stanley Eddington used to say that he wasn't sure he had never seen a star, in that he had seen only the light that comes *from* a star. Similarly, we might say that we have never seen a galaxy, in that we have seen no such single *thing* as a galaxy. What we see is light that informs us that a great many stars are there, and that they are arranged in a certain pattern. We call that arrangement a galaxy. But a galaxy is not a thing; it is a collection of things, or a collection of phenomena.

This is not meant to begrudge our view of the Andromeda Galaxy. It is alive with information and with heart-rending beauty. And it will improve as time goes by, for our galaxy and Andromeda, orbiting around a common center of gravity, are drawing closer together. Every second brings us fifty miles closer to Andromeda. In a few billion years, the two galaxies will be only half as far apart as they are today, and Andromeda will loom twice as large in our skies.

---

40 The Andromeda Galaxy (=M31=NGC224) at a distance of 2.2 million light-years is the nearest large spiral to our galaxy. Its colors result from a predominance of older red and yellow stars in the central regions that give way to a predominance of younger blue stars in the realm of the spiral arms. The stars scattered across the frame all belong to our galaxy and lie in the foreground, like raindrops on a window pane. Andromeda's two prominent satellite galaxies, analogous to the Magellanic Clouds that orbit our galaxy, are M32 (=NGC221), projected against an outer spiral arm, and NGC205 on the opposite side. If this were a photograph of the Milky Way taken from Andromeda, our sun would be located on the inner edge of one of the outermost visible spiral arms.

41 Viewed in detail, the inner region of the Andromeda Galaxy reveals delicate tendrils of dust and gas stretching for thousands of light-years. The white track across the upper left of the photograph is the trail of an earth satellite that passed while the exposure was being made (facing page).

42 The central regions of the Andromeda Galaxy appear in still greater detail in this black and white photograph. We cannot see a view of the central part of our own galaxy so clearly, for it is blocked from us by interstellar clouds (right).

43-45 The central regions (**45**) and outer spiral arms (**43** and **44**) of the Andromeda Galaxy are here resolved into individual stars. Though apparently numerous as grains of sand, the stars recorded on these photographs are only the brightest of the stars in Andromeda; more numerous dim stars like our sun lie beyond the threshold of these exposures.

42

43

44

45

46

48

47

49

80

50

46, 47 The two large satellite galaxies of Andromeda, M32 (top) and NGC205 (bottom), are composed primarily of old stars, unlike the Milky Way's satellite, the Small Magellanic Cloud, which contains many young stars as well.

48, 49 NGC147 (top) and NGC185 (bottom), two of the more remote satellites of the Andromeda Galaxy, are several hundred thousand light-years from their parent galaxy.

50 Galaxies emit energy not only in the form of light but in other wavelengths along the electromagnetic spectrum as well, among them those we call radio. This radio map, superimposed on a photograph of the Andromeda Galaxy, shows radio radiation coming from the nucleus and spiral arms of the galaxy.

The Galaxy M33

M33, a relatively diminutive Local Group spiral, is estimated to hold fewer than twenty billion stars. Its tiny nucleus contains less than two percent of the mass of the galaxy. Most of the rest of the mass is to be found in its spindly disk. The stellar population of M33 is heterogeneous, incorporating young blue stars, old red giants, and a wide variety of stars in between.

Radio studies of M33 have revealed that its outer arms are warped away from the plane of the galaxy, although this is not apparent in the photograph (right). The reason for the warp remains in question. No galaxies seem to be near enough to M33 to have produced the effect through recent gravitational interaction. Intergalactic clouds of cold hydrogen gas might have warped M33 through gravitational interaction, if the clouds were sufficiently massive. Radio astronomers have found evidence of such clouds nearby, and it may be that they have ruffled the composure of M33.

The Andromeda Galaxy must be a lovely sight in the skies of M33. It is only about seven hundred thousand light-years distant, and it presents M33 with a more open view of its spiral structure than the perspective granted us here in the Milky Way. If our solar system were located in M33, we would see the Andromeda Galaxy as a softly glowing beehive of light more than a dozen times the diameter of the full moon.

---

51 The spiral galaxy M33 (=NGC598), a member of the Local Group located near the Andromeda Galaxy, presents a relatively open face to our galaxy.

Figure 6. The Local Group
The galaxies of the Local Group are here plotted approximately to scale. The concentric circles mark intervals of one million light-years from the center of the Milky Way galaxy. Notice the pronounced binary nature of the Local Group: Most of its galaxies are clustered around its two dominant members, the Andromeda and Milky Way galaxies. The preponderance of satellites near the Milky Way is probably not a genuine effect, but reflects the fact that these dim satellites are more readily detected in nearby intergalactic space than are those at the distance of the Andromeda spiral. The Group is still far from being adequately mapped, and we may expect that many small Local Group galaxies remain to be discovered.

Maffei I

NGC147

NGC185

Andromeda
Galaxy
NGC205

M32

M33

IC1613

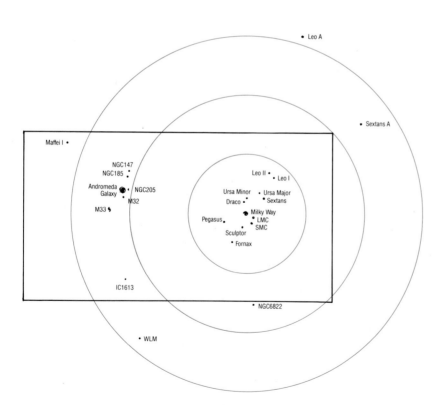

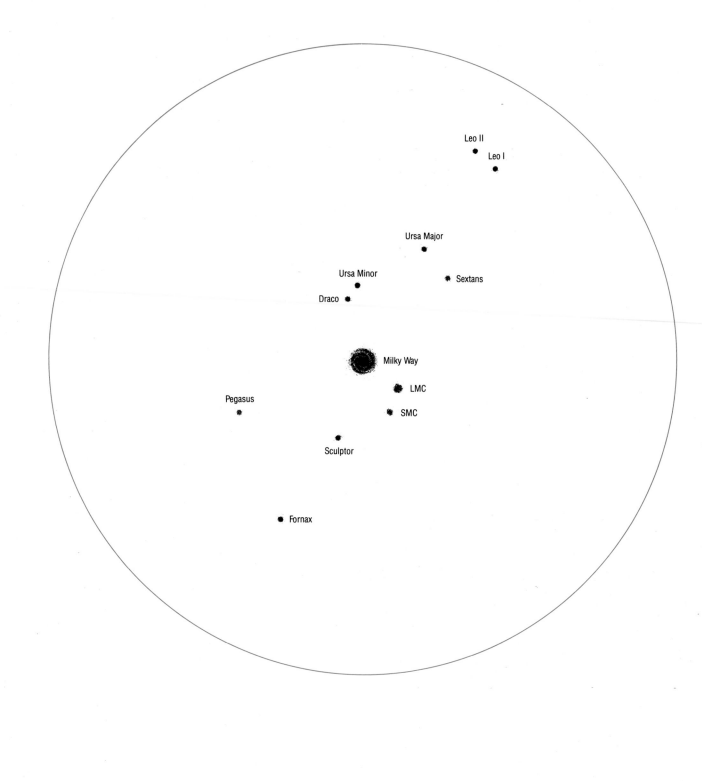

Leo II

Leo I

Ursa Major

Ursa Minor

Sextans

Draco

Milky Way

LMC

Pegasus

SMC

Sculptor

Fornax

The Sculptor Dwarf Galaxy

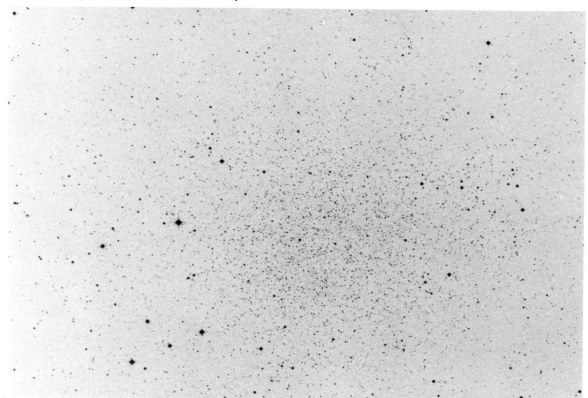

52 The sparse stars of the Sculptor Galaxy, a dwarf only about eight thousand light-years in diameter, are displayed here in a negative print that renders stars black and the sky white.

In the heavens as on earth we find plentiful evidence of what Shakespeare called "the modesty of nature," her predilection for the unassuming, the inconspicuous and the small. On earth we find more plankton than great fishes, more insects than human beings, dirt ubiquitous, gold rare. In our solar system, the most numerous objects are not giant planets like Jupiter and Saturn, or even modest planets like the earth, but the hoards of tiny planetlets called asteroids, few of them so large as a steamer trunk. In our galaxy, the majority of stars are dwarves. And among the galaxies themselves, the majority are not great systems like the Milky Way or Andromeda; they are the dwarf galaxies.

The Local Group contains dozens of dwarf galaxies. They are so paltry that altogether they contribute less than a tenth of the mass of the group as a whole.

The stars of the Sculptor Galaxy (above) add up to the equivalent of only about two million stars the size of the sun. The galaxy is so loosely organized that we can see right through it. It has no visible nucleus, and lacks the interstellar clouds characteristic of the spirals.

This deficiency of interstellar clouds may be due to the relatively low gravitational attraction of a galaxy as meager as Sculptor. Large galaxies are thought to garner interstellar material in two ways, by vacuuming up intergalactic gas as the galaxy moves along, and by the shedding of material by their stars in the form of stellar winds, "planetary" nebulae, novae and supernovae. But neither of these methods works well for a galaxy as small as Sculptor. Its gravitational force is insufficient to scoop up much gas from intergalactic clouds, some of which are more massive than Sculptor itself. And it is at a loss to retain much of the ejecta of its own stars; when, for example, a giant star in Sculptor explodes, the shell of its ejected material simply flies out into intergalactic space.

The Sculptor Galaxy is a satellite of the Milky Way, and orbits it in roughly the same path that is followed by the Magellanic Clouds and is demarked by the river of gas called the Magellanic Stream. There is evidence to suggest that a number of dwarf satellites of the Milky Way lie along a sort of great circle route defined by the Magellanic Stream. if so, our galaxy is being orbited by a chain of little galaxies.

III/The Form and Variety of Galaxies

*For those who are awake
the cosmos is one.*

—HERACLITUS

A Journey through Intergalactic Space

We accelerate out of the Local Group. As our speed edges ever closer to that of light, time on board passes ever more slowly by comparison to that of the universe at large. Seen through our eyes, the cosmos up ahead conducts its affairs with crazy haste. Planets whirl in their orbits. Stars are formed and die between breakfast and supper. We speed on.

We have entered upon the deep spaces that intervene between groups of galaxies. The familiar spirals of Andromeda and the Milky Way have shrunk until smaller than a fingernail at the distance of an outstretched hand. Free from nearby distractions, we are left for once in the sole and equitable company of everything—all the galaxies—floating in space in all directions. We while away our time by examining them through the ship's telescope.

What we see reminds us of home.

Back on earth, we recall, all the things of nature shared a deep kinship. Objects as dissimilar as snowflakes and stones proved to be made up of combinations of atoms drawn from a common pool of elements. Living things as diverse as a boll weevil and a human being were found not only to be made from the same common stock of atoms, but to have been built to the design of a single sort of molecule, that of DNA. All the creations of our planet could be understood as having been formed within the parameters of a few fundamental principles of physics. Yet for all their kinship, no two things could be found exactly alike—no two identical snowflakes or stones, boll weevils or people. Nature's way seemed to be to try everything without ever doing the same thing twice.

Now we find this way at work among the galaxies as well. All function within the purview of basic physical principles. Each galaxy, for example, must move through space along the trajectory dictated by its gravitational interaction with its neighbors and with the matter of the universe at large; no galaxy can pick up its skirts and scamper away in violation of those laws. And the material kinship of galaxies runs deep. All the stuff of all of them is made, so far as we can see, from various mixtures of the same sorts of atoms that we came to know back on earth.

Indeed, order and regularity are sufficiently manifest in the appearance of galaxies that we can sort them into categories.

About half the galaxies we see are spiral in form, like the

Milky Way. Some among our crew take chauvinistic pride in learning that their home galaxy is of the sort most widespread in the cosmos. Others more dispassionate point out that since most stars are to be found in spiral galaxies, the odds are that any given species evolving on a given planet circling a star will find itself in a spiral, as did we.

About one quarter of the prominent galaxies are ellipticals. Here the stars are arranged not in the flattened disk characteristic of spirals, but within a more nearly spherical volume of space. At first the ellipticals may look rather bland to our spiral-accustomed eyes, but as we study them further we come to appreciate their symmetry of form, their purity of content (ellipticals contain little interstellar gas and are made chiefly of stars and space), and the magnificence of their most exemplary representatives, which number among the largest galaxies in the universe.

Scattered among the many other galaxies we find the SO, or lenticular, galaxies, much like spirals in form but lacking spiral arms. They combine some of the qualities of both spirals and ellipticals.

A few percent of the major galaxies are irregular. Their virtues are those of individuality, even of eccentricity. In their splayed and contorted forms they offer us endlessly varied perspectives on the star fields and nebulae they contain, like translucent sea creatures, whose interiors and exteriors may at once be seen.

Dwarf galaxies abound, most of them ellipticals and irregulars. Frequently we find them ranged around larger galaxies. If they seem negligible by comparison, we need only consider that even a dwarf contains millions of stars.

If no two galaxies are identical, no two stars or planets identical, then how can we imagine the variety manifest in the universe on a planetary level? Is there to be found across the whole sweep of creation a single insect, flower, raindrop or mud puddle that somewhere has a twin? And where thoughtful life has arisen, to what degree do its imaginings converge with that of other intelligences, in consequence of nature's predilection for order and form, and to what degree do they diverge, in consequence of nature's predilection for variety?

What is the cosmology of imagination, we wonder as our imaginary ship wanders on.

III/THE FORM AND VARIETY OF GALAXIES

Normal Galaxies

SPIRALS

Most large galaxies are spirals. Their general anatomy can be described in terms of three components: a central region, elliptical in shape and centered upon the nucleus; a broad, flat disk shared by stars and interstellar clouds; and a spherical corona, or halo, composed primarily of old dwarf stars and globular clusters and embracing the galaxy as a whole. If we arrange spiral galaxies formally in terms of the size of the central bulge relative to the disk of each, we find that they can be placed along a continuum. A classification scheme based upon such an arrangement and widely in use among astronomers is illustrated on page 91. In this system, galaxies with large central regions are classified Sa, intermediates Sb, while spirals with small central regions relative to their disk are classified Sc and Sd. This one-dimensional scheme falls far short of accounting for all the important parameters of form in spiral galaxies, and debate continues about its details, but the fact that even a partially coherent system of classification is possible encourages the expectation that we may one day come to fully understand how matter came to be deployed across the cosmic theater in the form of galaxies.

Many elements of continuity have been discerned within the classification of spiral galaxies. The central regions, the bulges, of spirals are populated predominantly by old stars and are rare in interstellar material; in some ways they resemble elliptical galaxies like those pictured on pages 118 and 119. The disks of spirals are relatively rich in interstellar material; in many cases, new stars are being formed in the disk today, and as a result of this ongoing production of stars the stellar population of the disk is far more heterogeneous in terms of the ages of its stars than is the central region. Generally speaking, Sc galaxies with their relatively larger disks are more active in creating new stars, while Sa galaxies create stars more fitfully and less abundantly. The Milky Way Galaxy, an Sb or perhaps Sc located roughly midway between the extremes, we would expect to be populated by a heterogeneous mixture of old, middle-aged and young stars. And this is just what we do find here. Our sun is one of the middle-aged stars.

Some of the interstellar material in the disk of a spiral galaxy consists of dust that has been processed through the cores of stars that subsequently exploded, divesting themselves of material that included many of the heavier elements we call metals. The halo, composed primarily of stars that formed before there had been time for many of the heavy elements to be created, is generally metal-poor. The disk with its many younger stars is typically one hundred times richer in metals. A gradient in metal abundances may be detected across the disk itself: There are fewer metals in the more thinly populated outer disk—more metals in the interstellar inner spiral arms where the stellar population is dense and where more stars have been born and died. It should be remembered that these are generalities, like stating that St. Petersburg, Florida, has an elderly population (though not everyone in St. Petersburg is elderly), or that the Atlantic Ocean is salty (though the salinity of the ocean varies considerably from one spot to another).

The dynamics of spiral galaxies are elegant and subtle. Nothing illustrates this better than the spiral arms themselves.

We can readily see that spiral arms cannot be objects, like vines or tree branches. A spiral galaxy does not rotate all of a piece, like a phonograph record, but rather rotates differentially: Stars near the central regions orbit the galaxy much more rapidly than do stars in the outer reaches. In a typical spiral, disk stars near the central bulge complete an orbit in about twenty million years, while those in the outer precincts of the galaxy take some two hundred million years, ten times longer to complete one "galactic year." Much the same is true of the interstellar material of the disk; its rotation is likewise differential. If the spiral arms were all of a piece, differential rotation would quickly either fragment them or wind them tightly around the galactic center. The situation resembles that of runners confined to lanes on a track. Although they start abreast as they circle the track those on the inner lanes draw ahead. Soon the line of runners must either fragment as the outer runners fall behind, or, if the runners are to stay together, the outer ones must draw in toward the center until all are running on the inner track. Spiral galaxies have maintained their arms for billions

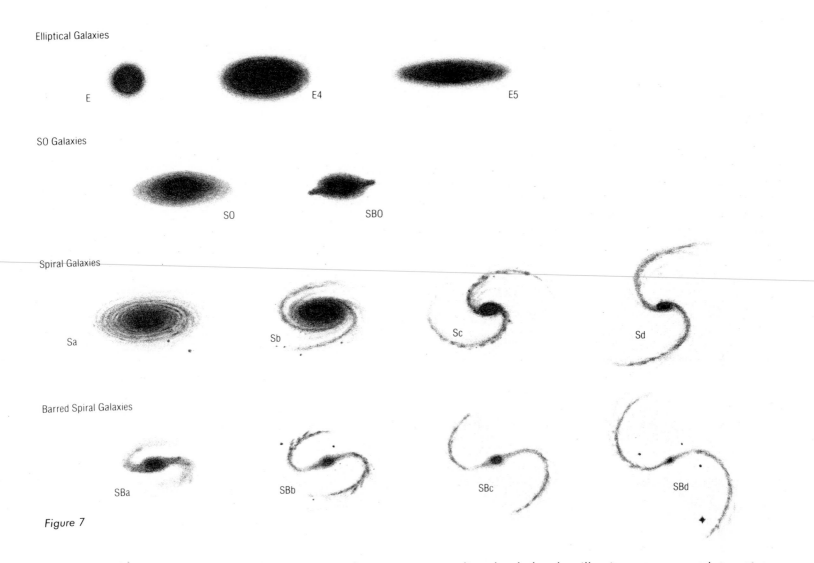

Elliptical Galaxies

E   E4   E5

S0 Galaxies

S0   SB0

Spiral Galaxies

Sa   Sb   Sc   Sd

Barred Spiral Galaxies

SBa   SBb   SBc   SBd

Figure 7

of years without either fragmenting them or wrapping them up, so we must abandon the view that they are physical entities and search for a mechanism that will explain them as a phenomenon.

The theories that have done this most successfully are those that view the arms as density waves propagated through the interstellar medium, like ripples in a pond. The density waves are set up by resonances in the gravitational interaction of the galaxy's billions of stars in their orbits.

We are able to see the spiral arms by virtue of the fact that when a wave encounters an interstellar cloud it often raises the density of portions of the cloud sufficiently for stars to form from the collapsing cloud. Their light, and that of the surrounding clouds that they illuminate, traces out the contours of where the density waves recently passed, and it is this luminous phenomenon that we see as the arms.

Figure 7. Galaxy Type Illustration
The formal continuum of galaxies is here illustrated schematically, with the spiral scheme extended to include galaxies with very small relative central regions classified Sd. Elliptical galaxies are classified E0 through E5 in accordance with their degree of flattening. Notice that perspective effects can greatly influence the classification of ellipticals, in that even a cigar-shaped E5 will look like an E0 if we happen to view it end-on. The S0 galaxies retain a category of their own, pending a better understanding of how they ought to be placed in the scheme of things.

68

The interstellar clouds of spiral galaxies tend to take on a disheveled appearance. The gravitational and magnetic fields of the galaxy tug at them, density waves compress them, starlight and stellar winds waft through them, and the clouds collide with one another.

Upwellings of interstellar clouds resulting from this interplay can be seen most evidently in galaxies that present themselves to us edge-on, as does NGC4565 (above). Two remarkable geysers, rising up from the plane of the galaxy, are silhouetted against the starlight of the central bulge. Off to the left a lofty pair of arches may be seen; these appear to consist of material that has been squirted up out of the disk and now, responding to the gravitational pull of the disk, is returning to it, like a spent skyrocket returning to earth. Celestial festoons like these may achieve altitudes of hundreds of light-years.

68 NGC4565 (above), tipped only four degrees from a perfect edge-on perspective, exhibits the components of a normal spiral galaxy—the elliptical central region and flat, dust and gas-laden disk.

BARRED SPIRALS

About one-third of all spiral galaxies are "barred" in form to a pronounced degree. By "bar" is meant a spindle-shaped grouping of stars and interstellar material that extends outward to either side of the central bulge and from which the spiral arms stem. Its length is typically a few tens of thousands of light-years. As the central bulges of many ordinary spirals seem to be somewhat oval or elongated in shape, it is possible that most spirals contain at least a vestigial bar. In the most widely accepted system of galaxy classification, the barred spirals are designated SB and are arranged according to the size of their central bulge along a continuum SBa to SBd, extending to irregular barred spirals like the Magellanic Clouds, designated SBirr.

Why some galaxies have bars is, like so many elementary questions about galaxies, still unanswered. One possibility is that the bar represents a way for the stars of a galaxy reared in disordered surroundings to settle into relatively stable orbits, stabilizing the galaxy in something like the way a tightrope walker regains his balance by extending his arms.

The barred spiral M83, one of the most dynamic-looking galaxies in the sky, looks almost as if it were tumbling, like a child's top kicked across the floor. In the absence of a photograph taken from another perspective by an observer in another quarter of the universe, we must try to decipher its three-dimensional shape as best we can from the vantage point granted us. So studied, M83 appears to have been warped, portions of its spiral arms bent well out of their original plane. Perhaps if we could see it edge-on it would resemble the "can opener" profile of NGC2146 (page 125).

Radio maps like the one on page 105 show M83 to be surrounded by enormous puddles of cold hydrogen gas. The form of these clouds suggests that of vestigial spiral arms lagging behind the arms of the galaxy within. This is just what we would expect to find had the galaxy indeed been warped relative to its surrounding envelope of gas. The galaxy would have gone on rotating, while the gas clouds, freed from some of the gravitational dominion that had been exerted upon them by mass concentrated along the plane of the galaxy, would have lagged behind. The gas cloud may

69

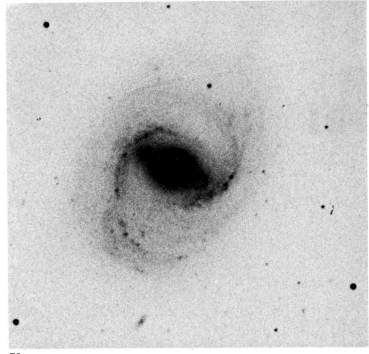

70

69, 70 The galaxies NGC3992 (**69**) and NGC4541 (**70**) are barred spirals; the "bar" is usually shaped something like a spindle, as the negative print of NGC4541 helps make clear.

101

71

72

73

71 The galaxy NGC1360 (left) is also a barred spiral.

72, 73 The barlike structures developed in many galaxies like these, NGC4650 (**72**) and NGC4548 (**73**), are thought to be dynamically stable forms into which stars may organize in a galaxy that has suffered from initial perturbations.

describe the old plane of M83, out of which the galaxy has since been wrenched and now sits like a barge listing in the sea.

What created the list? The only other galaxy in the vicinity of M83 is a small elliptical, NGC5253. It has only about one-tenth the mass of M83 and so ought not to have been able to precipitate so marked a disturbance, unless the two systems virtually collided.

LENTICULAR GALAXIES

Difficult to categorize are the SO, or lenticular (meaning lens-shaped), galaxies. Though shaped like spirals in that they have a central bulge and a large thin disk full of stars, unlike the spirals they have little interstellar dust and gas and no spiral arms. They might be described as elliptical galaxies cast in a spiral mold, and are sometimes said to represent an intermediate step between ellipticals and spirals. It may be more to the point to say that SOs look like nothing so much as spirals that have been robbed of their complement of interstellar gas and dust.

If so, what robbed them? According to one hypothesis some SOs are created when a normal spiral blunders into another galaxy or into a massive intergalactic gas cloud; such an encounter ought to leave the spiral in possession of its stars but swept clean of its interstellar material. Proponents of this explanation point out that SO galaxies are found most frequently in the inner regions of clusters of galaxies, where such collisions ought to occur most frequently. Opponents of this view point out that SO galaxies have been found drifting alone in free space, well away from any clusters; if intergalactic collisions were responsible for stripping their interstellar material, with what did they collide?

ELLIPTICAL GALAXIES

Elliptical galaxies present a simpler appearance than that of the spirals. In place of the mutli-component nature of a spiral galaxy—nucleus, central bulge, disk and corona—ellipticals are simply a case of billions of stars gathered in a roughly spheroidal volume. Most have not even a nucleus. The pristine clarity of the space between their stars is sullied by only the scarcest traces of interstellar material. A halo of globular clusters is customarily an elliptical galaxy's sole concession to ornament.

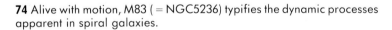

74 Alive with motion, M83 (= NGC5236) typifies the dynamic processes apparent in spiral galaxies.

75 The fanlike structure of interstellar clouds between the arms of M83 may be seen to advantage in this negative print.

76 Vast puddles of intergalactic hydrogen gas surround M83, here traced with a radio telescope at the twenty-one-centimeter wavelength.

75

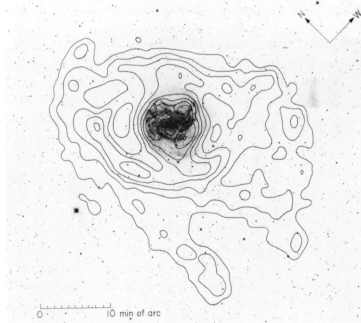

76

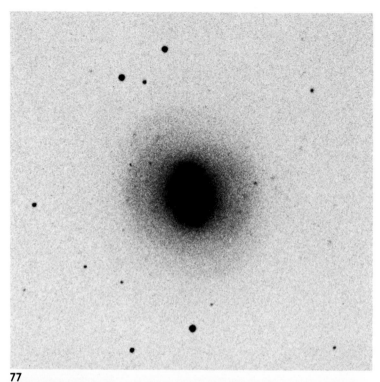

77

78

No perfectly spherical elliptical galaxy has ever been found, but some are nearly spherical. These are classified EO. Others, more flattened, take on shapes resembling squashed pincushions. These are categorized E1 through E5, according to their degree of flatness. The effects of perspective confuse the business of determining just how flattened an elliptical galaxy is, since even a radically flattened elliptical system will appear circular if we happen to be viewing it from the direction of one of its poles. As the astronomer Sir Fred Hoyle points out, an elliptical galaxy is always at least as flattened as it appears to be.

Unlike the stars of a spiral galaxy, which generally follow orbits that lie along the plane of the disk, like runners rounding a track, the stars of elliptical galaxies pursue orbits that are inclined at a great diversity of angles. Their orbits resemble the flights of hunting sea birds, some diving and then swooping upward while others circle variously amid them. A few elliptical galaxies are thought to be rotating as a whole; others display no evidence of rotation.

Spiral galaxies all lie within a relatively constrained range of mass, most having the equivalent of between ten billion and a few hundred billion stars like the sun, but ellipticals are far more varied in terms of mass. Dwarf ellipticals with only a few million stars and a diameter of but a few thousand light-years are common, while at the other end of the scale supergiant ellipticals have been found with populations estimated at ten thousand billion stars.

Unphotogenic, the ellipticals tend to be underrepresented in books of photographs like this one, though they constitute perhaps twenty percent of the prominent galaxies in the known universe. If we lived in an elliptical galaxy, however, we might cherish its Apollonian simplicity of form, so unlike the tangle of the spirals, and be grateful for the lack of interstellar clouds like those that block our view of much of the sky here within the Milky Way spiral. And we might take pride in a lengthy celestial history. Although ellipticals and spirals are estimated to be about the same age—roughly ten to fifteen billion years old—the ellipticals seem to have turned most of their raw material into stars quite early and thereafter gotten out of the star-making business. Today, the stars in

77, 78 The galaxy M84 (= NGC4374) (**78**), classified as an SO by some astronomers, as an elliptical by others, demonstrates the difficulty of galactic categorization. NGC4477 (**77**), on the other hand, is unambiguously an SO. Even when viewed in the detail of a negative print, its disk displays only the faintest traces of spiral arms.

ellipticals are predominantly old. They glow the dull orange of antique lamps, and that in a sense is what they are. Let us accord them the respect owed to elders. They were ablaze with the light of young stars, their planets were basking in that light, their stories were unfolding, when the earth and the sun were less than a whirlpool in a cloud.

IRREGULAR GALAXIES

Irregular galaxies introduce a touch of disorder into a cosmos otherwise dominated by the ethereal beauty of the spirals and the bald symmetry of the ellipticals. It is thought that they may be whipped into their disordered states in any of several ways. Many are satellites of larger galaxies; here it is clear that each might regain a classical form if it could spend some time away from the disruptive gravitational interference of the dominant galaxy. The Large Magellanic Cloud is an example; free from the Milky Way, it might be expected to reorganize itself into a more symmetrical form. Others may be transformed into irregulars by near collisions with passing galaxies. The disturbed appearance of NGC5195, recently disarrayed in an encounter with M51 (page 131), attests to this possibility. There may be other ways to create an irregular galaxy, but most irregulars seem to be little galaxies that have been bullied by bigger ones.

OUR PERSPECTIVE ON GALAXIES

The stars we see scattered across extragalactic photographs lie in the foreground and belong to our galaxy, not to the remote galaxies being photographed. We peer out at the universe through these scrims of stars, something as our remote ancestors viewed the world each morning from tree-branch perches through a foreground clustered with leaves.

The average density of the foreground stars varies considerably depending upon the part of the sky that intervenes when we train a telescope upon a particular galaxy. Foreground star fields generally are least dense where we are looking out at angles roughly perpendicular to the plane of our galaxy. The nearer our line of sight comes to the

79

80

79 M49 (= NGC4472) is an elliptical galaxy classified E4.

80 The Carina dwarf galaxy, difficult to distinguish from the many foreground stars in the Milky Way that lie along our line of view, is a dwarf elliptical.

81

82

83

84

81 NGC3077, an irregular galaxy, is gravitationally enslaved to the large spiral M81 (see page 136).

82 The Sextans irregular galaxy is an outpost member of the Local Group.

83, 84 The Sc galaxy NGC5364 (83) lies well away from the plane of the Milky Way, so that we see it with few foreground stars intervening, while the similar Sc galaxy NGC6744 (84) lies along a line of sight that passes closer to the Milky Way, where more foreground stars intervene.

85

galactic plane, the more foreground stars there are likely to be. When we try to look out along the plane of the galaxy —that is, through the Milky Way itself—we encounter so many stars that they all but clog the field of view, while massive interstellar clouds conspire to block our line of sight and obscure whole quarters of the universe from inspection at optical wavelengths.

NGC6744 (page 109) lies along a line of sight only twenty-six degrees from the plane of our galaxy, so we glimpse it through a robust thicket of stars. In contrast, NGC5364 (page

108) lies some sixty-three degrees from the plane of the Milky Way, and as a result relatively few foreground stars interrupt our view of it.

Galaxies are sometimes said to be "in" a constellation. Contemporary star charts divide the entire sky into constellations, their boundaries drawn so as to enclose the configurations of bright stars that our forebears named after gods, animals and other figures that caught their fancy or aided their memory. NGC6744 is in Pavo, the Peacock, a southern constellation catalogued in 1603 by the lawyer and

86

87

astrologer Johann Bayer. NGC5364 lies within the boundaries of Canes Venatici, the Hunting Dogs, an asterism once incorporated into the neighboring constellation Ursa Major but promoted to independent status by the seventeenth-century astronomer Johannes Hevelius of Danzig. Inasmuch as the stars of these constellations are close at hand compared to the enormous distances of the galaxies, galaxies may be said to be "in" the constellations only in the sense that the moon, seen through a window, may be said to be "in" the window.

85, 86, 87 The appearance of a spiral galaxy is influenced markedly by the angle at which we happen to view it. A spiral oriented face-on to us as is M74 (= NGC628) (**85**) affords us a view of the full articulation of its spiral arms. Nearly edge-on spirals such as NGC2683 (**86**) and NGC5907 (**87**) deny us much of a view of the structure of the arm, but recompense us by displaying something of the magnificent complexity of structure in the interstellar gas and dust lanes arrayed along the plane of the disk. M74 is classified Sc, NGC2683 as Sb, NGC5907 as Sc.

Supernovae, exploding stars of the highest order of violence, achieve such a brilliance as to be capable of capturing the attention of astronomers—such as there may be—throughout thousands of galaxies. One or two supernovae occur each century in a typical large spiral galaxy.

The light from each of these cataclysms rushes outward into intergalactic space, bringing news of the event to other galaxies with the speed of light. Observers in a galaxy five million light-years away will see the explosion after five million years have elapsed; those in a galaxy ten million light-years away will see it after ten million years. Therefore, the date that a given observer assigns to a given supernova depends upon his distance from it. In this, supernovae serve to remind us that intergalactic distances must be thought of in terms of time as well as space.

The supernovae marked by arrows in the two photographs

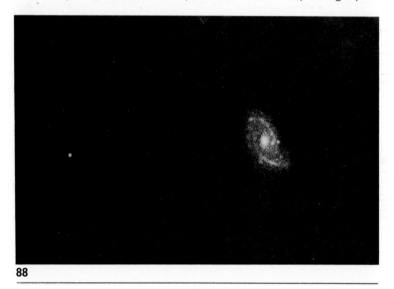

88

88 Extreme depth of field characterizes this photograph of the planet Pluto passing in the foreground of the galaxy NGC5248. When the photograph was taken, Pluto was a little under three billion miles, or 4.18 light-hours, distant. NGC5248 is some seventy million light-years away. The ratio between these distances is approximately the ratio between the diameter of the period printed at the end of this sentence and the distance from New York City to Sydney, Australia. Several dozen star images present on the original photograph have been removed from this print to facilitate identification of Pluto, a planet so small and remote that it appears virtually starlike in even the largest telescopes.

(page 113) were photographed here on earth in the same year, 1961. The galaxies in which they occurred, however, are different distances away from us. The spiral seen more nearly edge-on, NGC4096, is about forty million light-years away; light from its supernova had been traveling for forty million years when it reached earth. The more open spiral, NGC4303, is nearly one hundred million light-years away; the light from its supernova had been traveling for one hundred million years when it reached us. So we may say that in terms of a "cosmic" or universal time frame, the NGC4303 supernova must have occurred first, since its light has been moving through space much longer.

But this assertion could be disputed by astronomers in the galaxies involved. Imagine that there are astronomers in Galaxy A. Impressed by the brilliance of their local supernova, they base their calendars upon it, dating it as Year Zero. The light from this momentous supernova sets out on its trek across intergalactic space. Suppose that seventy million light-years separate Galaxy A from Galaxy B; therefore seventy million years must elapse before astronomers in Galaxy B can observe the Galaxy A supernova.

Before light from the supernova in Galaxy A reaches them, the astronomers of Galaxy B have seen a supernova in their own galaxy and similarly have chosen to date it as Year Zero. Ten million years later, light from the Galaxy A supernova reaches them. They record its date as Year Ten Million. The supernova log of Galaxy B therefore reads:

Galaxy B Supernova Log:
Galaxy B supernova, Year Zero
Galaxy A supernova, Year Ten Million

Now the light from the Galaxy B supernova is on its way to Galaxy A. By the time it arrives at Galaxy A, one hundred thirty million years have elapsed there—the sixty million years that had elapsed in "cosmic" time before the Galaxy B supernova occurred, plus the seventy million years required for the light from Galaxy B to reach Galaxy A. The Galaxy A log reads:

Galaxy A Supernova Log:
Galaxy A supernova, Year Zero
Galaxy B supernova, Year One Hundred Thirty Million

This is much the state of affairs with regard to the two supernovae pictured (page 113). Galaxy A is NGC4303, Galaxy B, NGC4096. Astronomers in both galaxies can claim with roughly equal justice that their local supernova was the earlier of the two.

89

90

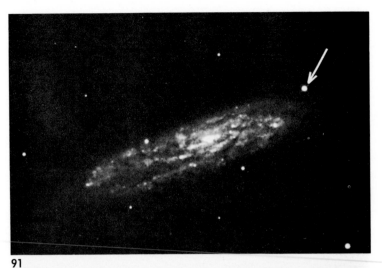

91

92

Here in the Milky Way, light from the two supernovae arrived at about the same time, and so we assign the same date to both—1961. As it happens, we have a local supernova of our own that remains fresh in cultural memory, the one that produced the Crab Nebula (see page 59), and we date it as having occurred earlier than either of these distant supernovae—in 1054 as opposed to 1961. But light from the Crab Nebula supernova has not yet reached either of the two external galaxies, NGC4303 or NGC4096, and will not reach them for millions of years to come. When it does it may be expected to show up on their logs as by far the most recent of the three.

In short, determining when things occurred in the cosmos is very much a matter of where they occurred relative to their observer. Changing our frame of reference will often reorder the temporal sequence we assign to events.

89, 90 The barred spiral NGC4725 is seen during and following a supernova. During its brief prominence, the exploding star shone nearly as brightly as all the billions of stars in the central regions of the galaxy.

91, 92 The exploding star marked by the arrow in NGC4096 (**91**) blew up sixty million years later than the one in NGC4303 (**92**) according to one scheme of reckoning time, but the light from the two explosions reached earth during the same year owing to the fact that the latter galaxy is sixty million light-years farther away than the former. Observers in each galaxy would date their local stellar explosion as the earliest.

Violent and Peculiar Galaxies

All galaxies emit energy—that is why we can see them—but some emit much more energy than do others. These are the "violent" galaxies, sometimes referred to as "active" or even "exploding." The energy output of an active galaxy can be prodigious, and may manifest itself not only in the wavelengths of the electromagnetic spectrum we call visible light, but elsewhere along the spectrum as well—at the longer wavelengths of radio and of infrared light and at the shorter wavelengths of ultraviolet light, X-rays and gamma rays. In

93

93 A longer exposure in black and white makes visible something of the extensive stellar corona of Centaurus A, a massive galaxy with perhaps three hundred billion stars.

94 Centaurus A, a powerful source of energy in optical, radio and other wavelengths, looks like an elliptical galaxy wrapped in the remnants of a spiral galaxy, and perhaps that is just what it is.

many cases, the locus of the galactic energy wellspring is that mysterious region, the nucleus.

The nuclei of active galaxies often show signs of hectic internal motion, their stars and interstellar clouds whirling about. They may be shrouded in thick dust, calling to mind a stop-action photograph of an explosion. And often these nuclei do appear to be exploding, in the sense that they are hurling matter out into intergalactic space.

The output of some of the more profligate violent galaxies would seem to require converting the equivalent of millions of stars like the sun into pure energy, in a cosmic crucible of unknown design. Where does this energy come from?

Adding to the mystery, the brightness of the nuclei of some violent galaxies varies. Some flicker over intervals as short as a few days or weeks; tremendous amounts of energy are being radiated here from something extremely small by galactic standards, perhaps only a light-week or so in diameter.

Various theoretical models have been put forth to explain what is happening in the nuclei of violent galaxies. It has been argued that their energy comes from collisions of stars, from a chain reaction of supernovae, or from a black hole that occupies the center of the galaxy, where it eats stars and interstellar clouds. But problems have arisen in trying to match any of these models to all the violent galaxies. For each violent galaxy that appears to fit a given model, another can be found that does not. One theory may explain how the nucleus produces so much energy, but fail to account for how it varies in brightness. Another may account for the variability, but not for the ejection of material from the nucleus into space. The galaxies in their wildly various behavior have something to tell us, if only we can learn how to better understand their language.

One promising clue lies in the very enormity of the violence. The fireworks of a violent galaxy are so spectacular that the galaxy cannot be expected to have sustained them indefinitely. To have done so, it would have had to convert all its matter into energy and have disappeared long ago.

This insight suggests that violent activity in galactic nuclei is a passing phenomenon, that it may be not a permanent attribute of a galaxy, like a person's having brown eyes, but a passing condition, like having the measles. To entertain this hypothesis requires that we consider the antiquity of the universe and the brevity of our tenure in it.

Imagine that you are looking out across a meadow full of

fireflies on a summer evening. Charmed by the sight of the fireflies, you make a snapshot of the meadow using an exposure of, say, one second. The resulting photograph proves to be something of a disappointment. Each firefly lights up for only a fraction of the time that it spends in flight. The rest of the time the firefly is dark, storing up energy for another flash. A one-second exposure will capture the light of only a fraction of the fireflies, those that happened to be flashing at the moment when the photo was made.

This may be our circumstance with regard to the violent galaxies. Perhaps many galaxies experience brief, recurrent episodes of violence, and the "violent" galaxies we see today are otherwise normal systems that happen to have been going through a violent phase at our epoch in cosmic history. They are fireflies that we have caught in the act of flashing, while the "normal" galaxies are fireflies biding their time between flashes.

The question remains, how do galactic fireflies flash? By what machinery can a galaxy's nucleus produce gales of energy, flinging the stuff of millions of stars into space with the abandon of passengers tossing streamers from the deck of a departing cruise ship?

Centaurus A (page 115), a giant galaxy with three times the stellar population of the Milky Way, produces great gouts of energy at many different wavelengths; alien astronomers whose eyes were fashioned to observe the world in X-rays or infrared light or radio waves would find it as commanding an object as we do in visible light. Most of these energies come from the region of the nucleus. In addition, radio energy is being generated by two pairs of clouds that lie along the galaxy's axis of rotation, that is, out from its north and south poles. Each of these pairs of radio-prominent clouds is symmetrical, one cloud on the north-pole side, the other on the south-pole side. The nearer pair lies sixteen thousand light-years to either side of the nucleus, while the other set is much farther out, at a distance of more than one million light-years from the nucleus. It is very possible that the clouds are composed of hot, thin gas that has been ejected from the nucleus.

The nucleus itself is variable, its radio and X-ray radiation altering in intensity over intervals of as little as a few days. Optically, the galaxy is very bright. At its distance of sixteen million light-years, it shines so brilliantly that it could be seen with the unaided eye in the skies of earth were it not for the thick swath of dust and gas that bisects it, cutting off our view of its central regions.

The dust ring earns Centaurus A the designation "peculiar," since normal elliptical galaxies contain little interstellar gas and dust. And while elliptical galaxies are dominated by old stars, as is the elliptical component of Centaurus A, its wreath is rich in bright blue stars, as can be seen in the photograph. The elliptical part of the galaxy glows around its edges with the red hue of older stars; in the more central regions starlight has overexposed the photograph so that colors cannot be seen there. Young blue stars are tangled through the wreath. Where we see the wreath superimposed on the inner galaxy, a coral color results from the light of the foreground blue stars mingled with that of the red stars behind.

The young stars of the wreath are very young, and must have been born near where we now find them, as none has had sufficient time to migrate far from its birthplace. The old stars of the elliptical realm are very old. Relatively few stars of intermediate age are to be found in this galaxy. It is as if Centaurus A were two galaxies merged into one. And perhaps this is a clue to solving the mystery of Centaurus A. Could it be that the elliptical component is a normal elliptical galaxy that recently swallowed a dust-laden spiral galaxy? In this scenario, the wreath is a remnant of the victim galaxy, the young stars in the wreath result from an episode of star-formation touched off by the shock of the galactic cannibalism, and the violence of the nucleus results from its having ingested interstellar material from the captured galaxy.

If Centaurus A did recently enhance its already imposing mass by absorbing another galaxy, most of the action was over by the time we came on the scene. Studies of the motions of stars in Centaurus A show that the elliptical component is rotating slowly, the wreath rotating more rapidly around it, but that otherwise the two components are drifting through space together. If a spiral galaxy was captured by Centaurus A, it has been captured for good and will never emerge.

Many of the known violent galaxies first came to the attention of astronomers using radio telescopes who noticed their unusually powerful output of energy at radio wavelengths. Radio radiation is common throughout the cosmos. Stars and even planets emit some energy at radio wavelengths, if only weakly, and normal galaxies keep up a soft radio murmur.

Most of the radio energy from a normal galaxy is produced by atoms of gas floating in interstellar space; each atom lets out a chirp of radio energy once in a great while; the number of atoms of gas in a galaxy is so large that a steady radio noise results, like the undercurrent of noise produced by a restless theater audience though only some of the people in the theater are speaking at any given time. This sort of radio noise is known as "line radiation," because each sort of atom radiates at a characteristic line, or frequency, along the radio spectrum. Hydrogen, by far the most abundant element in space, radiates at a wavelength of twenty-one centimeters; as a result, twenty-one-centimeter radio observations make for a useful way of mapping interstellar hydrogen clouds in our galaxy and in other galaxies nearby enough for this quiet but persistent form of energy to be detected.

In the far more powerful "radio" galaxies, the important source of radio noise is not the polite babble of atoms adrift, but the scream of electrons being accelerated to speeds approaching that of light. This means of energy production is known as "synchrotron" radiation, after the synchrotron accelerators used to speed subatomic particles in research laboratories. A violent galaxy may radiate one hundred times more powerfully in the radio wavelengths than does a normal galaxy, and some radio-prominent quasars (see page 177) are estimated to be pouring out energy yet a million times more powerfully still.

Cygnus A ranks as one of the most conspicuous radio sources in the sky despite its staggering distance of over half a billion light-years from earth. The radio noise comes from two lobes located symmetrically to either side of the galaxy, something like the twin radio lobes of Centaurus A. The similarity might go deeper, and the dark pinch that gives Cygnus A its hourglass appearance might be a dark wreath like the one that cuts across the face of Centaurus A.

Most radio-prominent galaxies are ellipticals, like Centaurus A and Cygnus A, but a few more nearly resemble spirals. Perseus A is radiating radio waves from a pair of lobes, but in this case the lobes are located close together near the nucleus of the galaxy , and are whirling about their common center of gravity so rapidly that they complete an orbit in only about ten thousand years. Each is estimated to be as massive as three hundred million suns. The intensity of radio output from this strange nuclear region varies acrobatically: Between 1960 and 1970 it boosted its output at the one-centimeter radio wavelength by more than five times.

The giant violent galaxy M87 heralds its presence by a bright jet that projects like a bony finger from its core. The jet is composed of hot, thin, ionized gas—what physicists call plasma—being shot from the center of the galaxy. It glows with an intense blue light produced by synchrotron radiation. This means of energy production, involving the interaction of

95,

96

electrons with a magnetic field, is usually encountered as a source of cosmic radio noise, but here in M87 the gas of the jet is being propelled through the galactic magnetic field with such violence that its energy has been shifted up from radio wavelengths into the more energetic wavelengths of visible light. We can get a sense of the velocity of the jet if we consider that although it is only some fifteen thousand years old it already has achieved a length of five thousand

95, 96 The galaxies Perseus A (**95**) and Cygnus A (**96**) are powerful emitters of energy at radio wavelengths.

light-years. By galactic standards it must have appeared as suddenly as a bolt of lightning.

The jet is emerging along one of the axes of rotation of M87—from the galaxy's "north" pole, if you will—and this axis is pointed somewhat in our direction. Evidence of a counterjet extending along the opposite pole has been found, but this jet is more difficult to observe since it lies toward the opposite side of the galaxy and is rushing away from us.

This situation, too, calls to mind Centaurus A, with its two pairs of radio sources oriented along the poles as if composed of clouds of gas that had been ejected from the nucleus. It may well be that in M87 we are seeing just such a classic two-lobe radio source in the process of creation. As the twin jets of M87, slowing and dissipating, coast on into space, we may expect that their energy level will drop back into radio wavelengths, lending M87 a radio profile similar to that of Centaurus A.

The dominant galaxy of the Virgo Cluster, M87 is enthroned at the cluster's center. A substantial amount of energy at X-ray wavelengths is being generated from intergalactic space in the environs of M87, apparently produced by clouds of hot hydrogen gas. If M87 is in the habit of spitting out jets of gas repeatedly, it could have produced these clouds in past eruptions. The astronomer Iosif Shklovskii estimates that one dose of gas spewed from M87 every three thousand years would be enough to account for the intergalactic gas clouds responsible for the X-ray radiation.

M87 is massive to be sure. Its more than three thousand billion stars would suffice to populate dozens of galaxies the size of our Milky Way, itself no minor system. The gravitational force exerted by this egg-shaped aggregation of stars and its more than ten thousand attendant globular clusters is appreciable by any standard. What resides at the center of it all, at the nucleus of M87?

97 The giant, violent elliptical galaxy M87 (= NGC4486 = 3C274 = Virgo A) with a mass of three thousand billion times that of the sun, a halo of ten thousand globular clusters, and a protruding jet composed of knots of gas, is one of the landmark galaxies of the universe, a conversation piece for even the most cosmologically urbane.

98, 99 In color, M87 displays the candlelight hues characteristic of the old stars that dominate elliptical galaxies, while the jet glows a blue-white resulting from synchrotron radiation produced by the interaction of electrons with the galactic magnetic field. The black and white photograph, underexposed to show detail in the nuclear region of M87, reveals that the jet proceeds directly from the nucleus.

100

101

A black hole?

The billions of stars located near the center of M87 appear to be orbiting at high velocity around a massive object that itself emits no light. This fits rather well with the hypothesis that the nucleus itself consists of a huge black hole that has consumed the mass of some five billion stars like our sun. Most of this diet, we may assume, consisted of clouds of gas, but the black hole might have augmented its diet by gobbling up whole stars as well. It is conceivable that the jets are made of scrap gas left over from the dismemberment of a star and slingshotted away from the realm of the black hole. Gravitational slingshotting is known to occur among stars, as when open star clusters lose some of their members in this fashion, and it is routinely employed to increase the velocity of interplanetary spacecraft when they soar past the giant planets Jupiter and Saturn. Slingshotting in the vicinity of a five billion solar mass black hole ought to be awesomely efficient. The mass of the jets we see today is about equal to that of the sun, a figure not inconsistent with the slingshotting scenario.

Much remains to be learned about M87 before the black hole nucleus can be considered more than just a likely hypothesis, but it does seem aesthetically appropriate to imagine that so much starlight is centered on an underworld composed of infinite darkness. If a Dante is to be found among the worlds of M87, here is a subject worthy of his attention.

The most violent of violent galaxies are the giant ellipticals like M87 and Centaurus A, but there are also many violent spiral galaxies. Their unusually high energy output characteristically comes from the nucleus. If we reflect that the central regions of spiral galaxies resemble small elliptical galaxies, we might say that a violent spiral is a spiral that happens to have a violent elliptical at its heart.

Some spirals with optically very bright nuclear regions are called Seyfert galaxies, after Carl Seyfert, who studied them in the early 1940s. Seyfert galaxies pictured in this book include NGC1275 (page 117) and NGC4151 (page 123), and these two galaxies, M77 (top) and NGC1566 (bottom).

The nucleus of M77 is a volcano of activity. Clouds of gas, each as massive as ten million suns, are being hurled outwards

---

100, 101 M77 (= NGC1068 = 3C71) (**100**) and NGC1566 (**101**) are two face-on specimens of spiral galaxies with unusually bright nuclear regions.

at speeds of nearly four hundred miles per second. The energy required to produce this maelstrom is equal to that generated by millions of supernovae. Conspicuous in visible light, the nucleus of M77 also radiates considerable energy at the wavelengths of infrared light and radio.

NGC1566 has a nucleus that varies in brightness as time goes by, flickering like a guttered candle. The distance from earth of NGC1566 is about the same as that of M77, seventy-five million light-years.

Many violent spiral galaxies show evidence of disorder along their disks as well as in their nuclear regions. In NGC4258 (above), the spiral arms terminate in bundles of bright new stars that stand out like the knots on the end of a cat-o'-nine tails. A pair of "ghost" arms, invisible but detectable with radio telescopes, trail behind the visible arms, a rather unusual circumstance. And instead of the smooth wheeling motion found in the interstellar clouds of normal spiral galaxies, M106 is wracked by what might be called interstellar storms

102 Signs of violent activity scar the spiral arms of NGC4258 (M106), a galaxy with an "exploding" nucleus.

that have sent clouds of gas scudding in wildly divergent directions.

The nucleus is unusually bright both optically and in radio wavelengths. There is evidence that it ejected two clouds of material with a total mass equal to several tens of million suns along the plane of the galaxy some eighteen million years ago. It seems fair to say that M106 recently "exploded," if we understand that term to mean not that the nucleus was destroyed, but that it violently disgorged itself of material in a manner somewhat analogous to a star's generating a "planetary" nebula—that is, in a traumatic but survivable episode, like a snake's shedding its skin.

At first glance M94 (above) looks more placid than its neighbor M106, with which it shares membership in the Canes Venatici

103 A brilliant nucleus marks the violent galaxy M94 (= NGC4736).

I Cloud of galaxies. But it too proves to have been the scene of an explosion, one that may have come as recently as ten million years ago.

The otherwise unremarkable appearance of some violent galaxies, like M94, bolsters the hypothesis that violent outbursts occur episodically in normal spirals. If so, what we call violent galaxies are really normal galaxies that we happen to see during one of their episodes of violent energy production. Perhaps all spiral galaxies know how to roar, and even such apparently quiescent ones as ours and the Andromeda spiral are but sleeping beasts.

Dominated by the fixed stare of its nucleus, NGC4151 is one of the more conspicuously violent galaxies in the sky. The brilliance of the nuclear region is so overweening that it makes it difficult to discern the structure of the rest of the galaxy; observing NGC4151 is something like standing in the path of an oncoming locomotive at night and trying to make out the shape of the locomotive behind its headlamp.

Observers who have managed to overcome this difficulty have found the structure of the galaxy to be that of a weakly barred spiral. The bar is not visible in the photograph (right), having been swamped by light from the nucleus. If we could "turn down" the nucleus we would see that the galaxy resembles NGC4156, the barred spiral that lies off to one side of it near the edge of the frame. (Although the two galaxies appear close together in the sky, NGC4156, at a distance of four hundred forty million light-years, is nearly seven times farther from earth.)

While the bar region of barred spirals normally is inhabited primarily by old red giant stars, the bar of NGC4151 shines most brightly not in red light but in blue and ultraviolet. Perhaps this blue light is coming from new stars that formed out of raw gas blasted from the nucleus of the galaxy into the spiral arms. The bars might be acting as conduits for material ejected from the nucleus, and might have become studded with new stars along the way in a manner analogous to the way mineral deposits build up on the inside of a water pipe—though the galactic bar is not, of course, a solid object like a pipe, but is rather an association of gas, dust and many stars.

The brightness of the nucleus of NGC4151 is variable. Soviet astronomers have identified what they believe to be a one-hundred-thirty-day period of variability for the nucleus, with a seventy-day pulse imposed upon it. One of the mysteries of violent galactic nuclei is how something that we would expect to be rather large manages to go through these gyrations, antic as a bee dance.

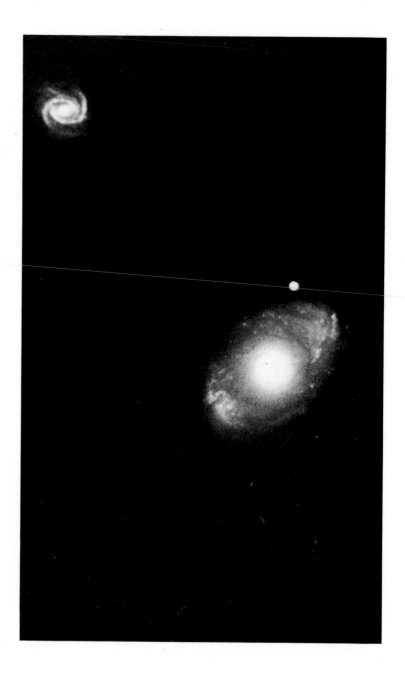

104 Obscured by the glare of its brilliant nucleus, NGC4151 possesses a barred structure similar to that of the background galaxy seen near the edge of the frame. Most of the foreground stars have been removed from this photograph so that the galaxy may be seen as in its natural habitat, floating in starless space.

If some galaxies may be said to be exploding, NGC253 may be said to be simmering. Clouds of gas are wafting outward from its central regions at a robust though not apocalyptic rate. Even so, the nucleus is losing gas rapidly enough that had it been simmering continuously throughout the history of the galaxy it would have depleted itself long ago. Since it has not done so, we may conclude either that the nucleus has acquired fresh material, perhaps by sucking in an intergalactic cloud, or that the outflow is intermittent.

A giant spiral, NGC253 is roughly the size of the Andromeda Galaxy. It belongs to the Sculptor Group, nearest neighbor to the Local Group, at a distance of a little over ten million light-years.

105, 106 The large spiral NGC253 shows only mild signs of activity in its nucleus.

107

108

"Peculiar" galaxies are those that fail to fit into a category under existing human schemes of galaxy classification. The term serves less as a description than as a convenient basket into which to dump galaxies our understanding of which is so limited that we cannot yet accomplish the zoologist's task of naming their family. Until we understand galaxies well enough to replace the term with a more learned one, we shall continue to enjoy the spectacle of humans calling galaxies "peculiar."

Many of the peculiar galaxies are violent, or are found to be conspicuously interacting with nearby galaxies, or both. NGC2146 is a good example; violently perturbed in appearance, it is a powerful source of radio energy.

NGC2685 arouses interest in that its structure is exceedingly unusual without there being anything else noticeably odd about it. The central component resembles an SO galaxy viewed nearly edge-on, wrapped in a set of gigantic hoops oriented perpendicular to its plane. A faint, barely visible outer ring encircles the entire system. Possibly we are seeing here the collision of two galaxies.

107, 108 These two strange-looking galaxies are classified "peculiar." NGC2146 (**107**) is classified SAb (pec), while NGC2685 (= Arp336) (**108**) is listed SO (pec).

IV/Interacting Galaxies

How is it that the sky feeds the stars?

—LUCRETIUS

A Journey between Interacting Galaxies

A high point in our intergalactic journey comes when we steer our ship between a pair of interacting galaxies. We have chosen to fly through the relatively narrow corridor separating two major spirals. They constitute a binary system, two galaxies bound together gravitationally as are the Milky Way and the Andromeda Galaxy. For most of their history they have stayed well apart, but now they are passing within only a couple of galactic diameters of each other, and it is at this dramatic stage in their interaction that we are to come between them.

The first mate is nervous. He points out that if we could see where we are headed in terms of Einstein's space-time continuum, we would perceive that our course lies along a precariously narrow ridge between two enormous wells created by the gravitational potential of the two galaxies. "We are steering between Scylla and Charybdis," he warns, "or rather Charybdis and Charybdis, for we'll have whirlpools to either side."

The galaxies are passing at an orientation that brings them almost face-on to each other, like a pair of cymbals. From a distance we see them edge-on. As the months pass and we draw closer, our perspective makes them appear to swing open like a pair of doors. The doors do not open evenly, but remain somewhat closer together near the top, where the relative inclination of the galaxies to each other has brought them closest together. Luminous tendrils bridge the gap between them there. Soon we will be amid the spectacle.

Thin clouds of hydrogen gas pervade the intergalactic space surrounding the two galaxies, and as we plunge through these their friction produces a sustained high-pitched wail from the ship's hull. To ward off nervousness we take solace in determinism: It is comforting to reflect that the cymbals cannot choose to clash when we pass between them, but will continue to follow the orbits dictated by Newton's and Einstein's laws. We have watched the projected course of the pair many times in computer simulations—a do-si-do, the two galaxies turning sharply around their common center of gravity and then parting a hundred million years from now. We ought to be able to sail between them without mishap. Still, we keep a cautious eye on our course; none among us wants us to attempt to be the first to fly at nearly the speed of light through a spiral galaxy edgewise.

To further comfort ourselves we discuss interacting galaxies in general. We remind ourselves that they are not rare. All galaxies, we reassure one another, may be said to be interacting, in that all respond to the general gravitational field of the universe, to which millions of galaxies contribute.

Didn't all the galaxies come from an undifferentiated soup of matter that permeated the universe long ago? And wasn't their formation a story of vortexes arising from the primordial soup, condensing to form the pairs, groups and clusters of galaxies we see today? And isn't the structure of galaxies, which we find so lovely a sight, but the visible message written by the invisible hand of these gravitational interactions?

There have been many close interactions of galaxies in the past, some perhaps involving the Milky Way and the Andromeda spiral, and the galaxies survived them in good order. They were merely twisted, their disks distended, their nuclei banked into fire, millions of their stars blasted into space.... Whole galaxies wrenched out of shape....

We fall silent. The ship's hull moans.

Ultimately we find ourselves between them. One spiral galaxy hangs to port, the other to starboard, two celestial wheels, ourselves at the axle. Their starlight flooding through the ports bathes the interior of our ship in a light such as none before us has known.

We view the spectacle from the overhead observation room, a transparent bubble that the ship's designers whimsically modeled after the domed railroad passenger cars once popular in North America. We turn out the lights and look above us to view the parts of the spirals where their mutual inclination has brought the disks closest together.

There the intergalactic gap is bridged by luminous tendrils that hang far above us like vines in an arbor. We can see that they are composed of gas and millions of stars being stripped from the lesser of the spiral galaxies and transferred to the more massive.

"A stellar caravan," remarks the first mate. "'The dogs bark; the caravan moves on.'"

We gag the first mate with an antimacassar, and resume watching, in silence, the transactions of galaxies.

The stars of a globular cluster flash past at close quarters, frightening us all. Amid screams, someone thinks to ungag the first mate. He bids us to be not afraid. Our course is taking us through the outskirts of one of the globular clusters that belongs to the halo of one of the galaxies, he shouts. Stars are flashing by the windows like balls from a Roman candle. He had intended to warn us, he shouts.

Still, we are quick to descend the ladder. It is days before anyone goes back up there.

Weeks pass and the twin galaxies crawl away aft. We welcome the sight of the dark intergalactic spaces we once had feared.

The M81—M82 System

112

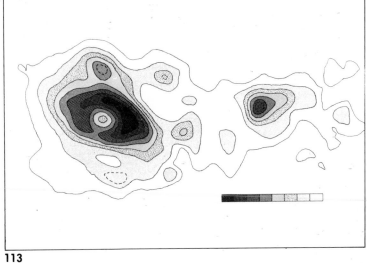

113

Another case of galactic interaction that occurred recently in cosmic history may be found in the M81 group, a neighbor of the Local Group only some ten million light-years away. The flagship of the group, M81 is nearly as large and populous as the Milky Way.

One hundred thousand light-years away from M81 lurks M82, a mysterious galaxy that has played the role of the Sphinx in contemporary astrophysics. It is probably a spiral viewed nearly edge-on; in any case it looks most odd. Its central regions are mottled in appearance, with dark clouds of remarkable extent silhouetted against massed starlight. Here are to be found bright nebulae in such abundance that if our solar system were located there the sky would be a glowing crazy quilt. Much of the interstellar material, dark and bright nebulae alike, extends far out of the plane of the disk, as if M82 had been shaken by some sort of galaxy-quake. Crowning its eccentric appearance are a pair of what appear to be clouds of hydrogen gas that protrude from the galactic poles.

A host of theories has been presented to account for what is going on within the shrouded confines of M82. Some of them are as imaginative as those that in times past promoted narwhales to the status of mermaids; these seem in retrospect to have resulted as much from the predispositions of the theorists as from evidence offered by the galaxy itself. Like

the Sphinx, M82 has conversed with us less in answers than in echoes.

Still it is possible to reconstruct a likely history of M82: Some two hundred million years ago it was churning peacefully through space, a small spiral going about its business. Then the enormous M81, ten times more massive than M82, swept past like an ocean liner past a sailboat. The gravitational wake of the larger galaxy washed across the smaller one, altering the orbits of millions of its stars and shocking its interstellar clouds into collapse, producing millions of new stars. Much additional interstellar material was jolted out of the plane of the galaxy, either tugged away by the pull of the receding M81 or blasted out of the disk by the many

---

111 M81 (= NGC3031) is the dominant spiral of a group of galaxies located only ten million light-years away.

112 A wide-angle view shows the relationship in the sky of M81, M82 (= NGC3034), and a dwarf satellite of M81, NGC3077, visible at the corner of the frame. Studies of the relative motions of the galaxies indicate that M81 and M82 passed each other two hundred million years ago and are now drawing apart.

113 A radio map of the system shows that M81 and M82 are wrapped within a common envelope of intergalactic gas.

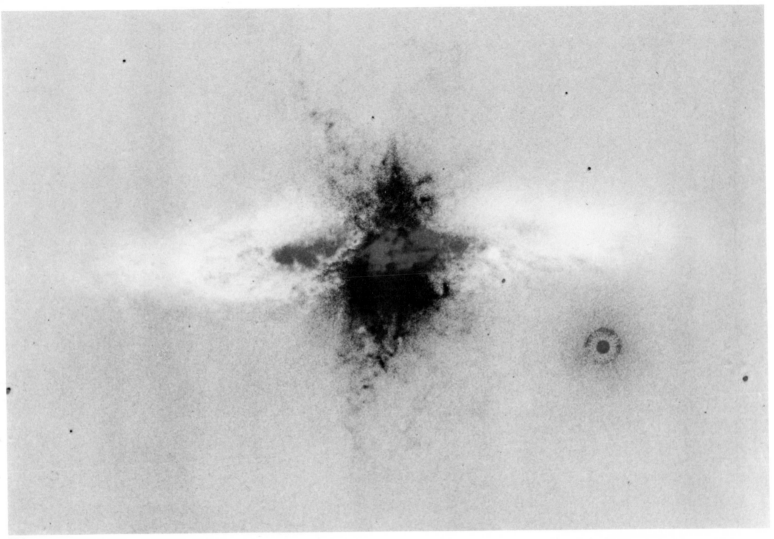

114

supernovae that characteristically flare up among young massive stars. The displaced material subsequently succumbed to the gravitational attraction of its parent galaxy and fell back into it, touching off another episode of star formation. Astrophysicists estimate that the first epoch of extraordinary star formation occurred primarily in the disk forty million years ago, and that the more recent one, in the central regions, is continuing today.

If this reconstruction is even approximately correct, then the M81—M82 system adds evidence in support of our dawning perception that some of the more spectacular events that occur in galaxies may result from their interactions with other galaxies.

---

114 Photographed in a wavelength of red light produced by hydrogen gas and designated Hydrogen Alpha, the clouds protruding from the central regions of M82 reach as much as ten thousand light-years into intergalactic space.

115 M82, highly disturbed in appearance, appears to be a small spiral galaxy seen nearly edge-on whose interstellar material has been shocked well out of the plane of the disk by the gravitational tug of the passing M81.

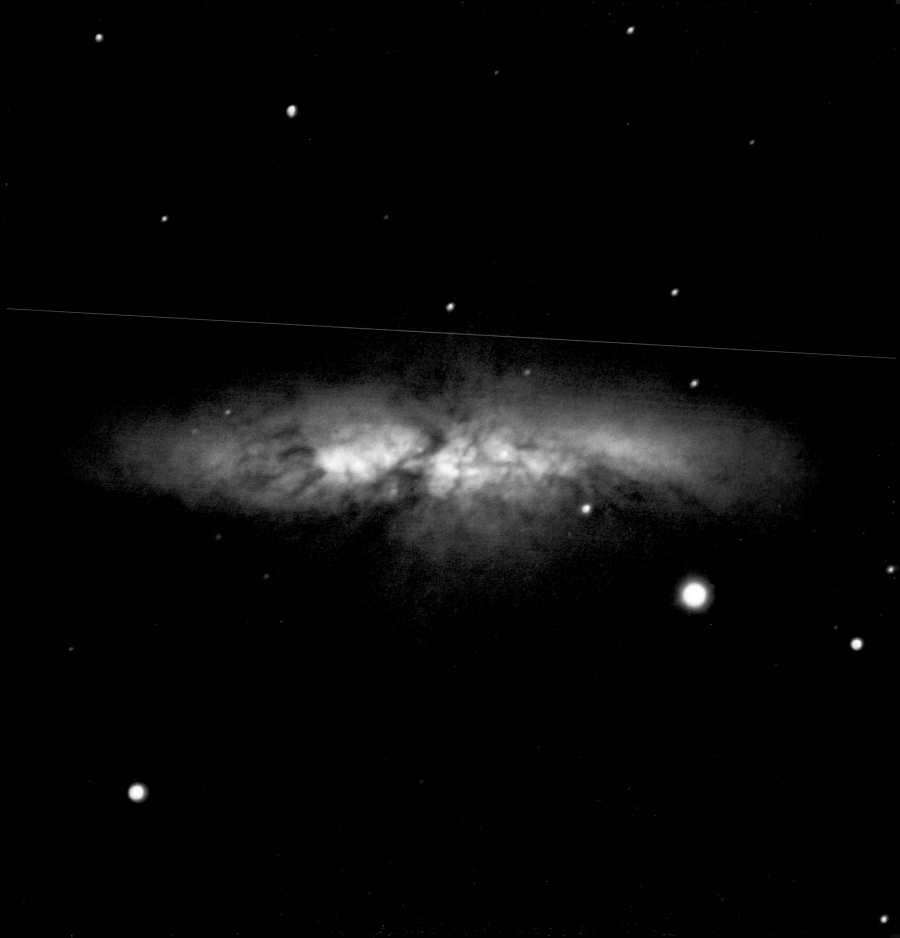

The Galaxy NGC4631

The giant spiral NGC4631 appears to be undulating like the sea in flood tide. The most likely explanation for its distorted appearance is that it is responding to the tidal pull of neighboring galaxies. It has two neighbors. One of them, NGC4627, may be seen in the photograph. It is small, pristine system—all stars and space—of the sort classified as a dwarf elliptical. The other companion, the spiral NGC4656, lies outside the field of the photograph. It has only about one-quarter as many stars as does NGC4631, but is currently only about one hundred thousand light-years away, less than the distance separating the Milky Way Galaxy from the

Magellanic Clouds. So its gravitational attraction ought to be sufficient for it to have distorted the larger spiral.

Since our perspective on NGC4631 is nearly edge-on, the interstellar clouds of its disk block our view of its nucleus in the wavelengths of visible light. At radio wavelengths, however, the nucleus can be detected, and is found to be highly energetic. Most of its radio energy comes from a bright central source and from a pair of secondary sources located to either side of it—a triplet pattern of radio emission found at the centers of many active galaxies. This suggests that the nucleus might be bright in optical wavelengths as well, could we see it through the intervening dust clouds.

Here again we may suspect that a mechanism that "turns on" the nucleus of a galaxy and makes it emit unusually large amounts of energy is triggered by the gravitational interference of other galaxies passing nearby. Might galaxies characteristically flare up when they encounter one another, their nuclear beacons flashing like ships signalling one another in the night?

116 The spiral galaxy NGC4631 is distorted along its plane, probably owing to gravitational interaction with two nearby galaxies.

117 The disk of NGC4631 glows in the blue color characteristic of young stars, while the copper-colored light of older stars of the central region, off-center in this distorted galaxy, peeps out from within.

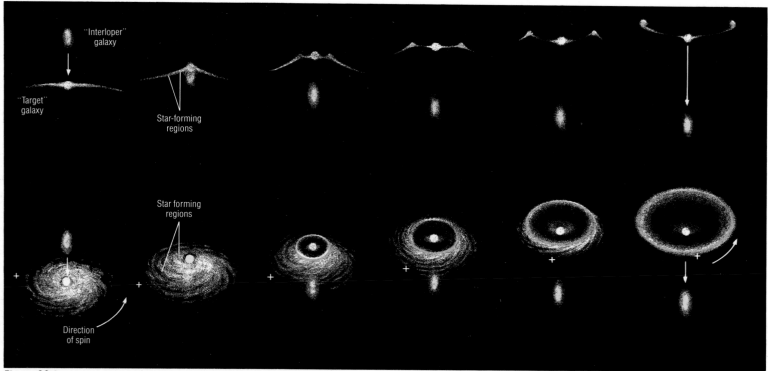

Figure 10

Ring Galaxies

Ring galaxies are created when a large spiral collides with a small galaxy or an intergalactic gas cloud. The ring structure, a temporary affair, results from the profound alteration of the gravitational field of the large galaxy that occurs when the interloper blunders into it.

Let us imagine that an interloper galaxy strikes a large spiral more or less dead center. Billions of interloper stars drift among those of the central region of the spiral, passing among them roughly at right angles to the plane of the larger galaxy. Interstellar space is roomy, and few if any stars collide, but the temporary occupation of the spiral galaxy by billions of alien stars has the effect of greatly enhancing the local gravitational attraction. Disk stars are drawn inward by the increased gravitation.

But by the time many of the stars arrive in the central regions, they find that the show is over. The intruder galaxy has passed on, taking its gravitational potential with it.

Released from the force that had attracted them, the disk stars rebound outward in an expanding ring. The shock of all this turmoil engenders the wholesale collapse of interstellar

---

Figure 10. Ring Galaxies
The creation of a ring galaxy is here reconstructed in two perspectives. A small "interloper" galaxy passes through or near the center of a large spiral, drawing the stars and interstellar material of the large spiral toward the interloper. When the interloper departs, the stars, gas and dust are released to fly back outward as a ring, spangled with the light of new stars set off by the shock of the event. This disfigured galaxy will soon regain its normal appearance.

123 An interloper galaxy speeds away, trailing a plume of gas, dust and stars and leaving behind a ring galaxy that already is beginning to reorganize itself into a normal spiral; the ring galaxy is catalogued as Object RG33 No. 754.

123

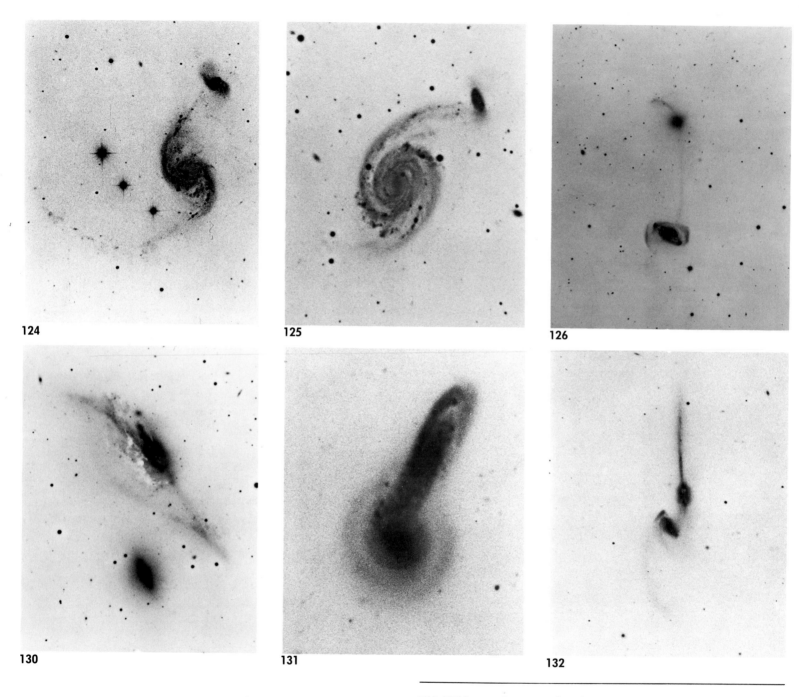

124

125

126

130

131

132

124–135 Interaction can give rise to an exquisite variety of form in galaxies. Here negative prints are employed, to show in greater detail in the rarified outer regions of each galaxy. They are: **124**: NGC2535/36 (= Arp 82); **125**: NGC7753/52 (= Arp 86); **126**: NGC5216/18 (= Arp

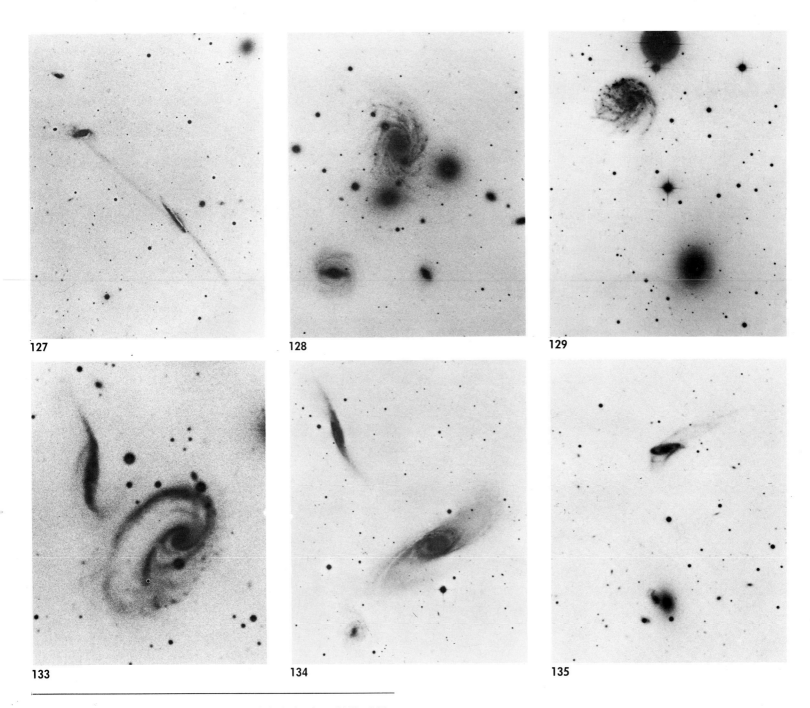

104); **127**: IC1505 (= Arp 295); **128**: NGC70 (= Arp 113); **129**: NGC2275/2300 (= Arp 114); **130**: NGC4438 (= Arp 120); **131**: NGC5544/45 (= Arp 199); **132**: NGC4676 (= Arp 242); **133**: Arp 273; **134**: NGC5566/60/69 (= Arp 286); **135**: NGC5221/22/26 (= Arp 288).

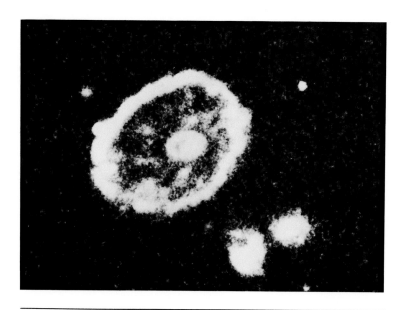

136 The Cartwheel is believed to be an otherwise normal spiral that has been violently disturbed by a direct collision with a companion galaxy, probably the upper of the two dwarf galaxies seen below and to the right of the Cartwheel.

clouds to form new stars. The expanding ring scintillates as it expands, twinkling with fresh starlight.

The spectacle is fleeting. The ring cannot long endure. Its stars sort themselves out into relaxed orbits and the galaxy resumes a normal appearance. Soon the only evidence of the collision is the spiral's unusually bright nucleus. Feeding on gas stripped from the intruder galaxy, the nucleus glows much like that of a Seyfert galaxy (see pages 117, 120, and 123).

The Cartwheel ring galaxy (left) may look insubstantial as a smoke ring, but it embraces as much space as the Milky Way, and harbors more stars. The companion galaxy believed to have collided with it is the slightly more distant of the two visible in the photo, the one that lacks a comma-like spiral hook. Measurements of its velocity lead to the conclusion that it passed through the center of the Cartwheel two hundred fifty million years ago. This agrees well with estimates of the age of the ring, adduced by measuring its rate of expansion and running the rate backward to the time when it would have been compressed near the nucleus, roughly three hundred million years ago. For almost every known ring galaxy a companion galaxy can be found nearby, slinking away from the scene of the collision.

V/Clusters of Galaxies

*In the universe the difficult things
are done as if they were easy.*

—LAO-TSU

A Journey through the Local Supercluster

Our old ship has gone far. We have edged so close to the speed of light that sometimes we feel we have become like light, fleet and insubstantial, velocity itself our only home. Decades have passed on board. There have been deaths and births, happiness and sadness, success and failure—in short, decades of life. The string quartet broke up years ago. The cook has grown grumpy from the ebbing of both praise and blame. Scholars complain about the limitations of the ship's vast library. We who set out on this journey when so young have become the elders. Occasionally we talk of putting in at a planet like earth, near a star like the sun, in a galaxy like the Milky Way, there to make a new start. But we are going so fast that just to decelerate would be the work of many years. So we fly on, like light.

Where previously we studied the form of galaxies, now our attention is drawn to the form of clusters of galaxies. Here we find order, intelligibility, a deep coherence underlying the diversity of the universe.

Clusters of galaxies, we see, display varieties of forms within a general pattern. The most straightforward way to arrange them is along a continuum in terms of structure, with the most regular clusters of galaxies toward one end and the most seemingly chaotic toward the other. The regular clusters are spherical or elliptical in shape, their galaxies concentrated at the center of the cluster. The irregular clusters, at the other end of our spectrum, are shambling and clumpy, often taking the form of extended chains of galaxies; they show little or no concentration toward the center. Intermediate between the two extremes are clusters that display some of the characteristics of both regular and irregular clusters; in some instances these consist of a central elliptical concentration surrounded by a halo or disk of more thinly distributed galaxies.

The forms of the clusters unavoidably call to mind the analogous forms of galaxies themselves: To some degree spherical clusters resemble spherical galaxies, irregular clusters resemble irregular galaxies, and intermediate clusters are not wholly unlike spiral galaxies, with their mixture of characteristics of both types. Our curiosity about this parallel deepens when we learn that the sort of galaxies predominant in each cluster is closely related to the form of the cluster itself. The spherical clusters have the largest plurality of elliptical and SO galaxies, while irregular clusters are dominated by spirals and have relatively few ellipticals and SOs. And what spiral galaxies there are in spherical clusters tend to be segregated toward the outer regions, or halo, of the cluster—much as globular star clusters occupy halos surrounding elliptical galaxies. The evidence seems compelling that the form taken on by a galaxy cannot have been determined solely by forces internal to that galaxy, but must reflect something of the milieu of the cluster to which it belongs.

Having taken this step up the hierarchical ladder, we are inclined to take an additional step and inquire whether clusters of galaxies belong to still larger associations. Here again our curiosity is rewarded. Many of the clusters prove to be members of superclusters—clusters of clusters of galaxies.

Clusters of galaxies typically occupy volumes of space with diameters of roughly thirty or forty million light-years. The diameters of superclusters are ten times greater, on the order of three hundred to four hundred million light-years. Even on this scale we find evidence of order and consistency. Some of the superclusters consist of a central zone where clusters of galaxies are relatively concentrated, surrounded by a flattened disk of more thinly distributed clusters, an arrangement at least faintly reminiscent of the structure of spiral galaxies. And possibly superclusters too are rotating.

Now when we look back to our home galaxy we may view it in a supergalactic context. The Local Group is a small cluster of galaxies located in the outskirts of the Local Supercluster. Many of our neighboring small clusters—the M81 Group, the M101 Group, the Sculptor Group—are likewise members of the Local Supercluster. The supercluster consists of a concentrated core, designated the Virgo Cluster, and an extended halo to which the Local Group and its neighboring groups belong.

A few of us gather in the observation dome after dinner. We trace for one another the structure of the Local Supercluster spread out before us, as once long ago we mapped the disk of our home galaxy. We talk of the old mystery that closes the circle of life—that the incomprehensible thing about nature, in Einstein's phrase, is our ability to comprehend it. However fast we go or far we travel, we have not fled one micron from this mystery. We can feel its breath, touch its face; it is our breath, our face.

The first mate stands and quotes a sentence from Carl Friedrich von Weizsäcker, a physicist and philosopher of science who lived millions of years ago on earth. "All our thinking about nature must necessarily move in circles or spirals; for we can only understand nature if we think about her, and we can only think because our brain is built in accordance with nature's laws."

The captain runs a hand through his white hair.

"Spirals," he says. "Our thinking expands as it circles. It moves in spirals."

V/CLUSTERS OF GALAXIES

The Form and Variety of Clusters and Superclusters

A cluster of galaxies may be defined as an association whose galaxies are bound together gravitationally. The shape of the orbit of each galaxy is determined by its gravitational environment in the cluster. In the case of a loosely organized cluster, the orbits of the galaxies might resemble the easy loops along which stars move in open star clusters, while in the denser spherical clusters of galaxies, the galactic orbits more nearly resemble the tighter orbits of stars in globular star clusters. In the Local Group, a small, loosely organized cluster of galaxies, the fundamental structure is binary—two galaxies, ours and the Andromeda Galaxy, constitute most of the mass of the cluster and orbit their common center of gravity.

Clusters of galaxies are found in a variety of forms that vaguely resemble the various forms of galaxies themselves. The so-called spherical clusters occupy a roughly spherical—elliptical would be a more accurate term—volume of space, their galaxies concentrated toward the core and scattered at the outskirts. This structure is reminiscent of globular star clusters and of elliptical galaxies, though the difference in scale is appreciable. If we were to represent a globular star cluster by a dot the size of a period on this page, a spherical galaxy would be a sphere seventeen feet in diameter, and a spherical cluster of galaxies would be the better part of a mile in diameter. Irregular clusters of galaxies, as their name implies, resemble giant irregular galaxies in their almost chaotic form. Many clusters occupy an intermediate position; in some cases these are found to consist of an elliptical aggregation at the center with a surrounding halo which, when flattened, resembles the structure of a spiral galaxy.

The size of regular and intermediate clusters of galaxies is normally a few tens of millions of light-years, while irregular clusters, like irregular galaxies, are much more varied in size, and include in their number many dwarfs. The Local Group, a few million light-years in diameter, is probably best classified as a dwarf irregular cluster of galaxies.

The types of galaxies found in clusters—and the vast majority of galaxies belong to clusters—reflect the structure of the cluster. Spherical clusters abound with ellipticals and their cousins, the SO galaxies. Irregular clusters are composed largely of spirals, with only a few ellipticals in residence. Intermediate clusters are populated by a mixture of types of galaxies.

This connection between the nature of clusters and the nature of the galaxies that inhabit them argues persuasively that the clusters are primordial, that they represent vast tracts of matter that were portioned out early in the history of the universe, before the galaxies began to form or at least before they had had time for the process of their formation to proceed very far. There must be a strong hereditary component—by which I mean such parameters as how much mass went into a protogalaxy, and what was the temperature and density of that mass—involved in the determination of whether a galaxy ends up as an elliptical, a spiral or whatever, and this hereditary influence ought to be traceable to conditions pertaining in the cluster in which each galaxy was born.

However there is also evidence for a strong environmental component acting upon galaxies throughout their lives. The experiences that befall a galaxy in the course of billions of years will differ considerably according to the nature of the cluster to which it belongs. A galaxy belonging to a loose irregular cluster follows a lazy orbit that only occasionally will bring it close to another major galaxy, and so it will have to endure galactic collisions rarely if at all. The environment is much different for a galaxy in a spherical cluster whose orbit plunges it into the densely populated cluster core. There it will be plucked at gravitationally by passing galaxies like a wealthy tourist pushing through a street of beggars, and collisions will befall it every billion years or so. A galaxy that passes directly through the center of a spherical cluster may find the conflicting gravitational pull of the thousands of galaxies surrounding it sufficient to tear it to pieces.

Supergiant galaxies are found at the center of many spherical clusters, and may well owe their opulence to their

Figure 11. Nearby Groups of Galaxies
Most galaxies belong to groups. Here are plotted a number of the groups of galaxies that have been identified in our part of the cosmos. The plane of the map is the "supergalactic plane" of the Local Supercluster of galaxies. The groups of galaxies, though typically irregular in form, are for simplicity depicted as spherical volumes of space. The numbers within each sphere indicate the distance of the cluster below (negative values) or above (positive values) the supergalactic plane. The concentric circles designate distances on the plane from the center of the Local Group. All these distances are in millions of light years, and all should be regarded as approximate.

Figure 11

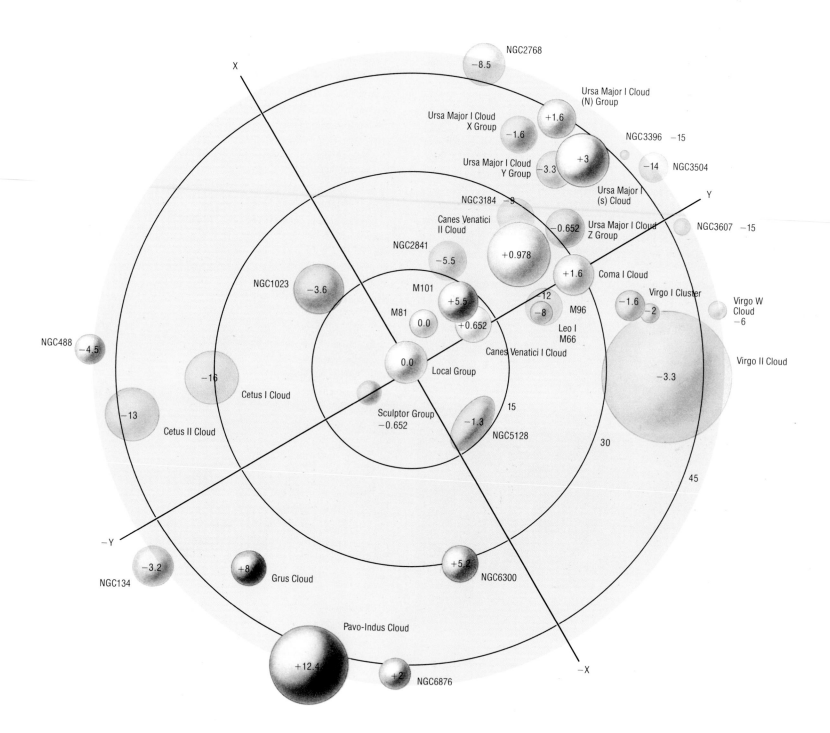

cannibalizing galaxies that blundered into the center of the cluster and fell victim to them. These supergiant core galaxies typically combine characteristics of the various sorts of galaxies they consumed—they may for example combine the shape of ellipticals with the rich interstellar clouds characteristic of spirals. And often they are found to have multiple nuclei, which may betray their predatory history as do the beaks of squid found in the bellies of whales.

Many clusters of galaxies in turn belong to superclusters, and the superclusters show evidence of structural order and variety analogous to that of clusters and galaxies. The fact that superclusters can be identified and that they bear the stamp of familiar physical laws offers reason to hope that one day we will be able, with reasonable accuracy, to reconstruct how they were formed and to predict their futures. So adept a level of comprehension might be expected to engender many a beautiful image in the mind's eye. For one thing, we might be able to correctly envision the dynamic behavior of superclusters with regard to the expansion of the universe.

"Expansion" is a way of saying that the clusters of galaxies are moving apart from one another at a velocity proportional to their distances apart from one another. Gravitationally bound, the clusters do not themselves expand. Rather, the expansion of the universe takes place out between the clusters. The clusters are departing from one another like swarms of bees that start out from one forest but diverge along differing paths of migration.

What about the superclusters? The expansion of the universe comes to predominate over local gravitational attraction on scales of something over three hundred million light-years, which is just about the average diameter of a supercluster. So it would seem that the gravitational interaction of superclusters must produce awesome distortions and transformations in supercluster structure as the ballooning of the universe tugs them apart.

When we can reconstruct the behavior of, say, the Local Supercluster over a period of billions of years, will we find that its motion more nearly resembles the orderly spin of a spiral galaxy or the gentle sway of sea grass in an ocean current? Or may its behavior, as is often the case in physics, evidence qualities that we had not anticipated, working more to engender fresh human metaphor than to lend itself to the metaphors of old?

The Virgo and Coma Clusters

The Virgo Cluster (right) lies at or near the center of the Local Supercluster, while our Local Group occupies a position toward the outskirts of the supercluster. It is perhaps seventy million light-years away in a direction roughly perpendicular to the plane of our galaxy. The photograph shows only one part of the cluster, which ranges across some twenty million light-years in all and is home to at least two hundred fifty large galaxies and perhaps a thousand or more lesser ones. Its population of many spiral galaxies mixed with a few ellipticals is characteristic of irregular clusters like Virgo.

In contrast to the Virgo Cluster, with its many spiral galaxies rather loosely dispersed, the Coma Cluster (left) is composed largely of ellipticals and the dust-free SO galaxies, relatively crowded at the cluster core. These are the characteristics of "spherical" clusters of galaxies. Over one thousand large galaxies and perhaps ten thousand dwarves are to be found here, brushing by one another at average leeways of less

138

than one million light-years—a density far greater than that of the Local Group.

The two giant galaxies visible in the photograph are a binary pair. Each in turn is orbited by a host of less massive galaxies. A binary structure of this sort is found in many clusters of galaxies. Evidence has been found that the Coma Cluster is part of an extensive, chainlike supercluster that may itself be binary; its companion, a cluster designated A1367, is two hundred fifty million light-years away.

137 Just under five hundred million light-years distant, the Coma Cluster of galaxies is home to an estimated thirteen hundred major galaxies, and is itself apparently subsumed within a super-cluster whose galactic population numbers over twenty-five hundred.

138 The Virgo Cluster of galaxies, seventy million light-years distant, is the nucleus of a supercluster that contains our Local Group and many other clusters of galaxies.

155

139

---

139 The Perseus Cluster of galaxies, three hundred and fifty million light-years distant, appears to us through a thick foreground scrim of stars in our own galaxy.

140 In this photograph of the Hercules Cluster (right), reproduced in negative to make visible many of the fainter galaxies, the sharply defined points and those with optically produced spikes are stars. Virtually all the other objects—all that are not obviously stars—are galaxies.

The Hercules and Perseus Clusters

Irregular in form, the Hercules Cluster lacks the central concentration of galaxies found in spherical clusters like Coma. But it has a noticeable structure, something like that of a meandering riverbed. Seven hundred million light-years away, it is one of four clusters belonging to the Hercules Supercluster, an association that stretches for a distance of fifty million light-years.

Many rich clusters of galaxies generate considerable amounts of energy in radio and X-ray wavelengths. Part of this energy comes of course from their member galaxies: Elliptical galaxies make up the most powerful X-ray and radio sources, and spherical clusters, where ellipticals abound, often show up prominently in X-ray or radio telescopes. But some rich clusters radiate X-ray energy from the regions between the galaxies as well. The source is thought to be clouds of hot gas ejected by elliptical galaxies or stripped from spirals that passed through the central regions of their clusters; it is from the central regions that most of the X-ray radiation comes.

In the case of the Perseus Cluster (left), about one-fifth of the X-ray output is generated by a single galaxy, NGC1275. Much of the rest of the X-ray radiation is emitted from an area nearly as wide as the cluster itself, some three million light-years in diameter. Additional evidence that intergalactic clouds are responsible is found when radio telescopes are trained on the Perseus Cluster. They reveal long "wakes" trailing behind several of the galaxies in the cluster, as if the galaxies were churning through intergalactic gas like boats cutting through water.

VI/Galaxies and the Universe

...It is no one dreame that
can please these all....

—BEN JONSON

A Journey toward the Edge of the Universe

Now comes an end to our journeying. Millions of years have passed on the planet of our birth, decades aboard. Clusters of galaxies pass abeam and are recorded in the logbooks as once we recorded the passing of stars and later of galaxies. The time has come to turn the ship around and decelerate until it can be brought to rest on a planet. We owe this much to the younger generations, who never saw the earth and have known only this life of ceaseless exploration. But for us elders it is the beginning of the end. Deceleration will take a long time, and we cannot hope to live to see the day when our crew will step out onto planetary soil, under planetary skies.

On the day when the deceleration order is to be given, we few survivors of the original crew gather in the observation dome for a last look at the cosmos while our ship is at its peak velocity. Few visit the observation dome any longer—to swim among the galaxies is unremarkable to those who have known no other surroundings—but to our old eyes the view remains awesome and a little frightening. The time-dilation effect having sped up the workings of the spiral galaxies that lie ahead of us by a factor of several million, they spangle with the light of millions of newly formed stars, and still more brilliant supernovae flash and crackle across them by the hundreds.

The captain rises with difficulty and proposes a toast. "To the unattainable goal," he calls, his glass raised toward the dome and to the galaxies in array. "To the edge of the universe."

"Hear, hear," we respond. How often we have talked about the edge of the universe, mapped it with our telescopes, saluted it with this same toast. The phenomenon is as familiar to us as our names.

When we look across space we are also looking back in time. At distances of up to a few billion light-years we see galaxies as they were recently in cosmic history, looking much like those that lie nearby. At distances of five to ten billion light-years we are seeing younger galaxies whose light set out on its journey when the universe was about half its present age. At distances approaching fifteen billion light-years what we see are the brilliant beacons of galaxies being formed; they pour out huge quantities of energy, by comparison to which the births of stars in the contemporary cosmos seem but a bland reminiscence, like firecrackers set off to celebrate the anniversary of a revolution. At these distances we are seeing the denizens of a young cosmos, all light and noise.

The captain, projecting into cosmic time the tendency of the old to aggrandize the historical, sometimes speaks of events fifteen billion years ago as if he had been alive then, rather than having only witnessed them vicariously by telescope. "That was when galaxies were galaxies," he likes to say. "The juice squeezed out of ten thousand stars in a year. Stars blowing up with every tick of the clock. Energy aplenty—you could singe your hair just by stepping outdoors—and galaxies crowded so close together there was scarcely room to pass between them. A pilot had to keep on his toes in those days."

If we search with our telescopes for galaxies more distant than those at some fifteen billion light-years, we see nothing. At these distances we are looking back to a time before the primordial stuff of the universe had cooled sufficiently to congeal into stars and galaxies. That is what we mean be the edge of the universe—a temporal threshold marking the point in cosmic history before which darkness prevailed. It is an edge not of space but of time, and to visit it we would need not our spaceship, but a timeship able to travel into the past.

"To the unattainable goal." The first mate echoes the toast. "Faster than the galaxies."

This is a traditional riposte, one that refers to the expansion of the universe. The farther away a given galaxy we observe, the faster it is receding from us (or us from it, as you prefer) as it takes part in the universal expansion. In all directions we see galaxies on the threshold of the universe hurtling away from us, trains out of the past running on rails of the past, their lights the markers of the unattainable past.

The captain orders the ship brought about.

"Faster than the galaxies," the first mate repeats. "Fast as light." A mathematical witticism contained in special relativity prescribes that the fuel bill to accelerate any particle of matter to the speed of light would be infinite, would include the conversion of everything, itself included, into energy.

"Maybe the young folks are right to want to stop," the captain says. "They probably figure that otherwise we'd go on forever, that we'd burn up the whole universe in order to cross it. They figure we'd be firing up the boilers with tables and chairs when we'd left not one star shining in the sky."

"Don't worry, Captain," says the mate. "There will always be space travelers."

"Always have been," the captain replies. "We were space travelers before we ever left earth. See that galaxy over there?" He extends a gaunt finger. "When they look at the Milky Way, don't they see it speeding away at ten percent the speed of light? And that galaxy over there, don't they see it moving off at twenty percent the speed of light? And those millions of galaxies off near the edge, don't they see our galaxy moving almost as fast as light itself, just as we see them? Aren't we teetering on the edge of the universe, so far as they're concerned? Isn't it our part of the universe that's young and blinding bright, so says the old light that left here so long ago and only now is reaching them?

"We are all space travelers, gentlemen. We are. They are. All are."

The galaxies wheel across the sky as the ship is turned end-for-end.

"Let us show a light," says the captain. He produces a kerosene-burning ship's lantern, a treasured antique. He lights the wick, replaces the glass, and holds the brass lantern up to the windows of the dome. Its yellow flame mingles with the light of the galaxies.

"In a moment this flame will belong to our past," he says. "But it belongs to their future. Maybe one day an astronomer in one of those galaxies whose telescope is pointed the right way at the right time will catch this flicker from our little lantern. Just a couple of million miles' worth of light falling into the telescope, gone in a couple of seconds."

The captain blows out the lantern, sets it on the floor, takes the con, and gives the order to fire the engines.

"It's not so bad to be old, gentlemen," he says. "We're part of the future of most of the universe."

VI/GALAXIES AND THE UNIVERSE

Geometries of Space and Time

Albert Einstein once remarked, in a rare display of impatience over those who complain that the theory of relativity violates common sense, that "common sense" for each of us consists of what we learned prior to age sixteen. If we wish to improve our understanding of the cosmos we would do well to heed Einstein's remark, laying aside the prejudices of our common-sensical comprehension of things and taking to heart rules appropriate to the interstellar realm.

Prejudices dating from the childhood of our species run deep. For most of our history, we humans have tended to regard the earth as both stationary and central to the cosmos. This was an illusion. The earth is not stationary, but is moving in orbit about the sun, the sun in orbit about the center of the Milky Way Galaxy, the galaxy around the center of gravity of the Local Group, the Local Group in its orbit within the Local Supercluster, and the supercluster is moving as it participates in the expansion of the universe. Nor do we inhabit the center of the universe; no one does.

The cosmos is a study in motion and change. To set up a frame of reference in such conditions we are well advised to pay attention not only to *where* things are but *when*. The location of, let us say, the Rock of Gibraltar can be specified adequately using only the three dimensions of space, so long as our frame of reference is constrained to the surface of the earth: We simply say that it is located at the mouth of the Mediterranean, at about thirty-six degrees north by five degrees west. But these specifications no longer suffice once we step away from the earth and take a broader view. If we take neighboring stars as our reference point, the Rock of Gibraltar is being whisked along by both the rotational and orbital velocity of the earth; if we step back further and take an intergalactic perspective we must add in the sun's galactocentric velocity, and so forth.

In order to specify the location of Gibraltar in anything greater than a parochial context, we must therefore specify its location not only in the three dimensions of space but also in the fourth dimension of time. Relativity may be viewed as an attempt, and a highly successful one, to build this dictum into the foundations of physics. It does so by viewing events in a context of where-when called the space-continuum.

Four-dimensional geometry can present conceptual problems for creatures as visually oriented as we are, in that four-dimensional structures—a 4-D sphere, say, or a 4-D cube—are difficult if not impossible for us to visualize. A 4-D geometer must work without the aid of the mind's eye, like a perfumer imagining a fragrance yet to be created or Beethoven composing music from within the chambers of his deafness. These obstacles aside, viewing the universe in terms of a four-dimensional space-time continuum has made possible highly encouraging progress in cosmology, the science concerned with discerning the structure of the universe as a whole.

To see how the addition of a fourth dimension can deepen and refine our concepts of the cosmos, let us examine how adding a dimension can solve cosmological problems that arise on lower orders of dimensionality. Specifically, let us examine the dilemma of whether the universe is finite or infinite. How might this dilemma be resolved by creatures whose perceptions are limited to one-dimensional or two-dimensional frames of reference?

Imagine a one-dimensional world. Its citizens are known as Linelanders, a name that I have borrowed, like that of the Flatlanders to follow, from Edwin Abbott's lovely book *Flatland*. The actions and perceptions of Linelanders are confined to only two directions, forward and back. Each lives his life in a perpetual queue, just behind one Linelander and just in front of another. Passing muttered words up and down the queue, the Linelanders debate whether their cosmos is finite or infinite in extent. They know perfectly well that the *inhabited* world is limited; there is a Linelander who stands at the head of the line, and another who stands at its end. But beyond them the Line stretches into the distance. Does it ever end?

Some Linelander cosmologists maintain that it does not, that the Line is infinite in length. Their favorite form of argument is the reduction to absurdity. If the Line were finite, they argue, what would happen when one came to the end of it? How can there be an end to the cosmos? The notion is unthinkable; therefore the Line—the cosmos—must be infinite.

But Linelanders who favor the concept of a finite cosmos can point to equally troubling absurdities in the infinite cosmos model. One of their favorite lines of argument goes like this: If the Line is infinitely long, then where is its center? Why, nowhere. Or everywhere. Indeed, there is no universally agreeable way to designate points of reference on an infinite line. Yet here we are, standing on a given point, and there is another point on the Line a foot away; for this to be the case must not the Line be finite? One old Linelander cosmologist likes to refute the infinite-universe cosmologies by scoring the Line with a chalk mark and echoing Samuel Johnson's refutation of Berkeley: "I refute it thus!"

This is the dilemma of the finite versus the infinite universe. It was an ancient ponderable when Lucretius wrote about it in the first century B.C. And it appears to be a genuine dilemma, insoluble within the scope of dimensionality available to the intuitive apprehensions of the Linelanders.

Lineland

Finite but Unbounded Lineland

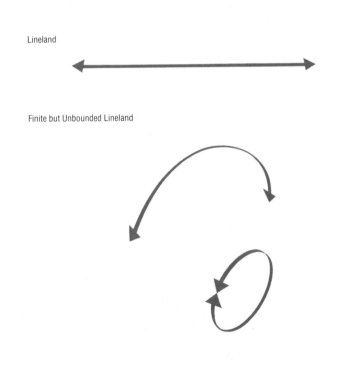

Figure 12. A Finite but Unbounded Lineland
The inhabitants of Lineland, limited in motion and perception to forward and backward, contemplate the dilemma of whether the cosmos is finite or infinite. An infinitely long Line seems unthinkable, but no less troubling is the question of how, if the Line is finite, it comes to an end. By introducing a second dimension—up and down—a Lineland cosmos may be created that though finite in extent has no boundary or edge. The finite-infinite dilemma is transcended.

But the finite-infinite dilemma for the Linelanders can be transcended if we add a dimension, permitting the Lineland cosmos to be bent.

Suppose we bend it into a loop. The result is that almost magic creation, a finite but unbounded universe. The Line has a finite length, yet it never comes to an end (Figure 12). A squad of Lineland explorers can be dispatched in any direction—remember, they have only two to choose from—and travel as far as they like without ever encountering an edge to their universe. If they travel far enough, they will find themselves back where they started. Their report reads: The universe is finite but unbounded; it has no edge and our choice of what to call its center is arbitrary. The debate is over and the cosmologists are free to turn their attention to deeper questions.

A similar transformation can be wrought upon the cosmos of two-dimensional creatures (Abbott's Flatlanders) by providing them with recourse to a three-dimensional cosmology. The Flatlanders live in a world of forward and back, right and left. So long as they confine their thinking to these two dimensions, they face a similar cosmological dilemma to that confronted by the Linelanders—either the plane is infinite in extent, or, equally unthinkable, it somewhere comes to an end. The dilemma can be evaded by introducing a higher dimension. Suppose we wrap the plane inhabited by the Flatlanders into the three-dimensional form of a sphere. Voila! The Flatlanders too have been bequeathed a finite but unbounded universe (Figure 13).

The Flatlanders, limited in their perception to two dimensions, cannot directly intuit the fact that they inhabit the surface of a sphere. But they can discover the shape of their spherical cosmos by means of experiment. If they dispatch an expedition of explorers to circumnavigate the globe, that feat will constitute convincing evidence that the Flatlander cosmos is spherical. A more subtle experiment can be performed by Flatland geometers without leaving home: They may lay out a triangle on the ground and measure the sum of its angles. Inflated by the swell of the globe, the angles will add up to more than the 180 degrees of a flat triangle. So Flatlander cosmologists, though unable to intuit or perhaps even to imagine a sphere, can deduce the spherical cosmos mathematically and intellectually.

We who think of ourselves as living in a three-dimensional world can accomplish similar feats by adding a fourth dimension to our conception of the cosmos. Working in four-dimensional geometries we can construct many models of our cosmos that are finite, in that they are composed of a

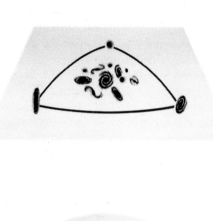

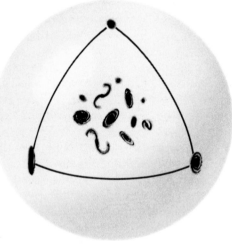

Figure 13. A Finite but Unbounded Flatland

For two-dimensional as for one-dimensional creatures, the introduction of a higher dimension overcomes the dilemma of a finite versus an infinite cosmos. Here a two-dimensional figure, a plane, is wrapped into the three-dimensional figure of a sphere, again resulting in a finite but unbounded universe.

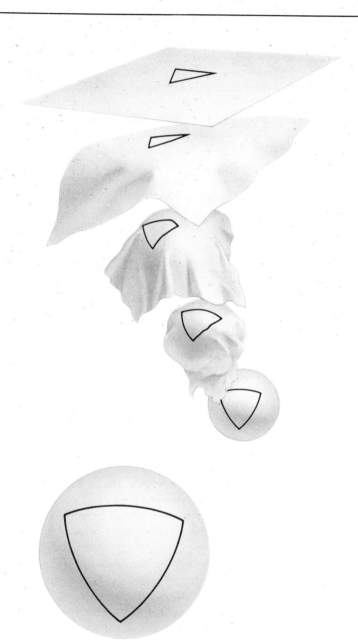

Figure 14. Spacetime Geometry Traced by Means of Light Beams

Lines that appear curved when viewed within the constraints of a given level of dimensionality may prove to be straight when viewed in terms of a geometry incorporating an additional dimension. In the two-dimensional universe represented schematically in the diagram at the top, a dense concentration of galaxies has affected the trajectory of light rays passing between galaxies A, B and C so that they appear curved to two-dimensional observers. But when a third dimension is added and the surface of the diagram is wrapped into a sphere (bottom), the same light rays are seen to be traveling by the shortest distance between A, B and C. In relativity, light rays that seem curved as observed by us three-dimensional observers actually are following the shortest path available along the contour of the four-dimensional space-time continuum.

finite number of galaxies and a finite amount of space separating them, and unbounded, in that one can travel in any direction indefinitely without reaching an edge. Many four-dimensional shapes of the cosmos may be imagined, but for the sake of simplicity let us stick to a spherical model and examine what might be some of the characteristics of a four-dimensional spherical universe.

To trace the geometry of a four-dimensional universe, let us use beams of starlight. This approach seems sensible since we can hope for no straighter line than a light beam, and it has the historical virtue of having been a favorite method of surveyors since the ancient Egyptians helped to invent geometry by employing sight lines to survey the boundaries of farms on flood plains of the Nile. Like Flatlanders tracing a triangle on the surface of their spherical cosmos, we can examine the four-dimensional structure of our universe by examining beams of starlight across large distances.

When we do this, we find that the light beams are not perfectly straight. They bend, and the degree to which they bend is directly related to their proximity to matter. A beam of light passing near a star curves toward the star and departs on an altered trajectory. Such a curvature of the space-time continuum was predicted by the general theory of relativity, and as it happened it was an observation of the bending of beams of starlight near the sun that constituted the first experimental proof of the theory.

Having established that light beams curve in the universe, we might say that space itself is curved. This is not an unreasonable statement, but to insist upon it is to needlessly assert the parochialism of our three-demensional prejudices. To see why, we rejoin the Flatlanders.

The Flatlanders who traced out a triangle on the surface of their spherical world might call the sides of that triangle curved, but we who can perceive the nature of a sphere see that they in fact represent geodesics, or the shortest distance between the points that they connect. International air travelers are familiar with this effect. A plot of the shortest air route for a round trip, say, from Los Angeles to Tokyo to Auckland and back to Los Angeles, perfectly efficient if plotted on a globe, will look curved if plotted on a flat map.

So it is theoretically possible to plot the shape of the cosmos in four dimensions by mapping the course of light beams passing across distances that make up an appreciable segment of the dimensions of the cosmos as a whole. Such distances are vast, and direct measurements of universal geometry in this fashion lie as yet beyond our reach, so that it

141

remains to be learned to which of the possible four-dimensional forms the contours of the universe most nearly conform.

Fortunately, we, like the Flatlanders, are not restricted to a single avenue of cosmological experiment, and can investigate universal geometries by other, less direct means. Since the fourth dimension of our four-dimensional paradigm is time, we may hope to learn something of the geometry of the universe by inquiring into its behavior over long periods of time.

---

141 This cluster of galaxies, barely visible among the bright foreground stars, is about five billion light-years distant. It represents the outer limits at which galaxies can be optically observed with existing telescopes and photographic equipment.

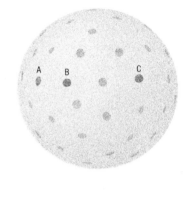

Figure 15. *Expansion of the Universe*
An expanding universe is here portrayed in terms of Flatlander cities
arrayed upon the surface of a sphere. At each stage from left to right
the size of the sphere has doubled, and so therefore have the distances
between any two given cities. Since cities that were farther apart to start
with must have traveled at higher relative velocities to double their
distance in a given period of time, expansion reads out to any observer
as a state in which the more distant a given city, the faster it is seen to be
receding. By studying this phenomenon, Flatlanders can infer that they
live in an expanding universe, as we infer expansion from the universal
recession of galaxies. Note that for the Flatlanders the universe is not
expanding *into* anything; it is simply stretching. The same may be said
of our universe, which may be conceived of as a finite but unbounded
system within which the density of the matter comprising the universe is
decreasing as time goes by.

The Expansion of the Universe

Once Einstein had created, in relativity, a means of interpreting
the cosmos in terms of four-dimensional geometry, he learned
that his theory strongly implied that the universe was either
expanding or contracting. Relativity would not, it seemed,
permit the universe to remain static, but required that it either
be opening like a blossoming flower or closing like a wilting
one. This implication startled observers and theorists alike,
Einstein not least among them. It had long been understood
that there was plenty of motion in the cosmos, but the idea
that the cosmos as a whole was engaged in *coherent* motion
was radical. At first few took it seriously. Then astronomers
discovered that remote galaxies are rushing apart from one
another, and us from them, at velocities directly proportional
to their distance—in short, that the universe is expanding.

To examine what is meant by an expanding universe, let us
make a last visit to the world of Flatlanders on a sphere
(Figure 15). Dotted across the surface of the sphere are
Flatlander cities; these may be taken to represent clusters of

galaxies in the universe. Now imagine that the sphere is
being inflated.

Three intriguing characteristics of an expanding universe
can be demonstrated using this model.

First, Flatlanders residing in any given city find that every
other city is rushing away from them. This is true for every
Flatlander, no matter in which city he lives. Each may choose
to regard his city as being at rest, or another city as being at
rest and his in motion, or all in motion; the choice is arbitrary.

Second, the rate at which the cities are rushing apart is
directly proportional to their distances. Consider three cities,
A, B and C. At the onset of our period of observation, A and
B are one hundred miles apart, while B and C are two
hundred miles apart. One unit of time later, the sphere has
doubled in size. All intercity distances have doubled as well.
A and B are now two hundred miles apart, B and C four
hundred miles apart. We can see readily that for B and C to
have opened up twice as much intervening space between

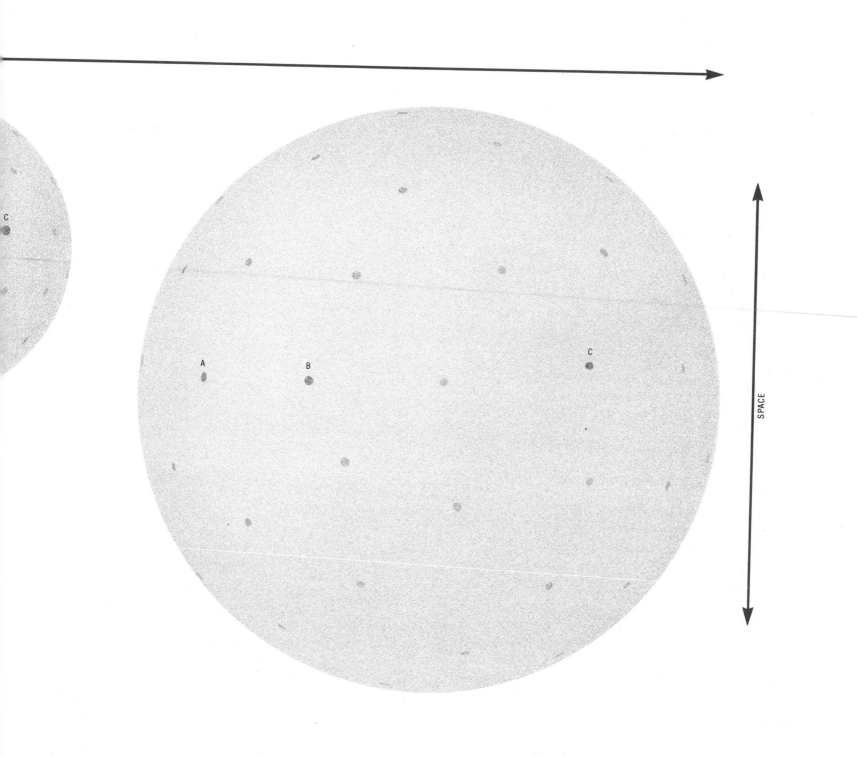

168

them as did A and B in the same period of time, the velocity of B relative to C must have been twice that of A relative to B. And the relative velocities of A and C, the two cities in our sample most distant from each other, must be still higher. That is the nature of expansion. Observers in any given Flatlander city can infer that their universe is expanding once they discover that all the cities are rushing apart at velocities directly proportional to their relative distances. It was by discovering this fact about clusters of galaxies that humans learned of the expansion of the universe.

A third characteristic of our expansion model is that no observer occupies a privileged or central position. No Flatlander city lies closer to the edge of the universe than any other, and none lies closer to the center of the universe, for on the surface of a sphere is to be found neither edge nor center. All observers are equally able to discover evidence of the expansion and to derive similar conclusions from it, and all observers see cities in all directions. Cosmologists call such a universe isotropic, meaning that its general appearance and behavior appear the same to all regardless of their location. This again seems to parallel the situation pertaining in the universe of galaxies: Every observer sees galaxies receding no matter in what quarter of the sky he looks.

It remains to be said that the shape of our universe is not necessarily, or even probably, analogous to a four-dimensional sphere. It may prove to be so, in which case it belongs to a class of "closed" geometrical forms, or it may adhere more closely to one of the "open" four-dimensional forms. If the universe is destined to stop expanding one day in the future, it is said to be closed, in other words to conform to a geometry analogous to that of a sphere. If instead the expansion of the universe is destined to go on forever, its geometry is said to be "open," or more nearly analogous to one of the various hyperbolic four-dimensional forms, usually described as saddle-shaped. Or the cosmos may prove to fit more closely to the template of a more exotic geometric figure, perhaps one involving still higher orders of dimensionality.

Since one of the dimensions of the space-time continuum is time, it is possible to theorize about the shape of the continuum by reasoning from information about how the expansion of the universe has proceeded during the course of cosmic history. The expansion of the universe, if engendered in a violent "big bang," ought not to have proceeded in a wholly unbridled fashion. The gravitational pull exerted by the clusters of galaxies upon one another ought to have retarded the rate of expansion to some degree. Cosmological models may be constructed in which the geometry of the cosmos is deduced from the deceleration rate.

How then do we investigate the fate of the universe? An ideal way would be to peer into the past. If we could observe how rapidly the universe was expanding billions of years ago, we could then compare that rate with measurements made locally of its rate of expansion today, and from the difference between the two determine the deceleration rate and predict whether or not expansion will go on forever.

Here, as in so many other ways, the universe is happily accommodating. Intergalactic space is so transparent that we can see galaxies billions of light-years away. And since light coming from those great distances has taken billions of years to reach us, we see the universe there as it was long ago. It is possible to study cosmic history directly. This effect the astronomers call "lookback time."

Lookback Time

What we see in the sky is the past. Light falling upon the earth tonight from the star Sirius, 8.7 light-years away, is 8.7 years old. Light from the red star Antares, 520 light-years away, dates from the fifteenth century. We see the Andromeda galaxy as it was in the first days of *Homo Erectus*, the galaxies of the Virgo Cluster as they were when coconut palms grew at the North Pole and terror cranes darkened the skies of Earth. Light from distant quasars set out on its journey to our telescopes before the earth had formed. To look across space is to look back in time. The history of the cosmos is arrayed in the sky for those who care to read it.

Some of the implications of this situation may be investigated by means of a diagram in which space is plotted as the vertical axis and time the horizontal (Figure 16). The "light

---

142 How many galaxies are there? No human yet knows. A rough estimate, based upon counting galaxies in our galactic neighborhood and extrapolating for the universe as a whole, is one hundred billion. This map plots the location of one million galaxies, or roughly one in ten thousand of all those in the universe. Its lacy, membranous pattern represents the largest-scale structure yet glimpsed by the human eye.

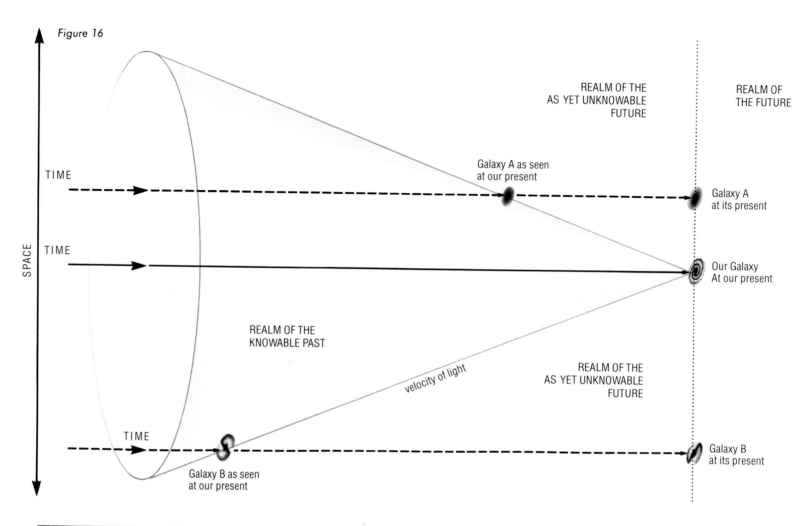

Figure 16

REALM OF THE
AS YET UNKNOWABLE
FUTURE

REALM OF
THE FUTURE

TIME

Galaxy A as seen
at our present

Galaxy A
at its present

SPACE

TIME

Our Galaxy
At our present

REALM OF THE
KNOWABLE PAST

REALM OF THE
AS YET UNKNOWABLE
FUTURE

velocity of light

TIME

Galaxy B
at its present

Galaxy B as seen
at our present

Figure 16. Intergalactic Past and Future Viewed in Terms of a Light Cone
Events occuring elsewhere in the cosmos make themselves known to us only when their light (or other radiation) reaches us. Therefore, events may be divided into those that lie within our "light cone," that is, events whose light has had time to reach us, and those lying outside the light cone, about which we can as yet have no knowledge. Our galaxy may be envisioned as moving from left to right in this diagram as time passes, so that events are constantly being swept into our light cone.

As the light cone diagram illustrates, the more distant the galaxy the farther back in time we see it. Galaxy A, relatively nearby in space, appears as it was in the recent past when its time line intersected the boundaries of our light cone. Events that transpired in Galaxy A since now belong to the history of Galaxy A, but lie in our future since they have not yet entered our light cone. Galaxy B, more distant, intersects our light cone at a point farther back in time and so is seen by us as it was longer ago.

cone" in the illustration is created by drawing its sides along a slope equal to the velocity of light, time against distance. At any given point in cosmic history, the events that may be known to a given observer are limited to those inside his light cone (Figure 16).

If the universe were static and unchanging, the fact that we can see back into its history would be of little use to us. But as we live in a changing, evolving universe, lookback time

Figure 17. Light Cones of Observers in Three Galaxies
Here our galaxy and Galaxy A see each other in the relatively recent past, while Galaxies A and B, farther apart, see each other in the more remote past. Quite recent events in each galaxy are as yet unknown to observers in other galaxies, the light announcing them not yet having had time to traverse intergalactic space.

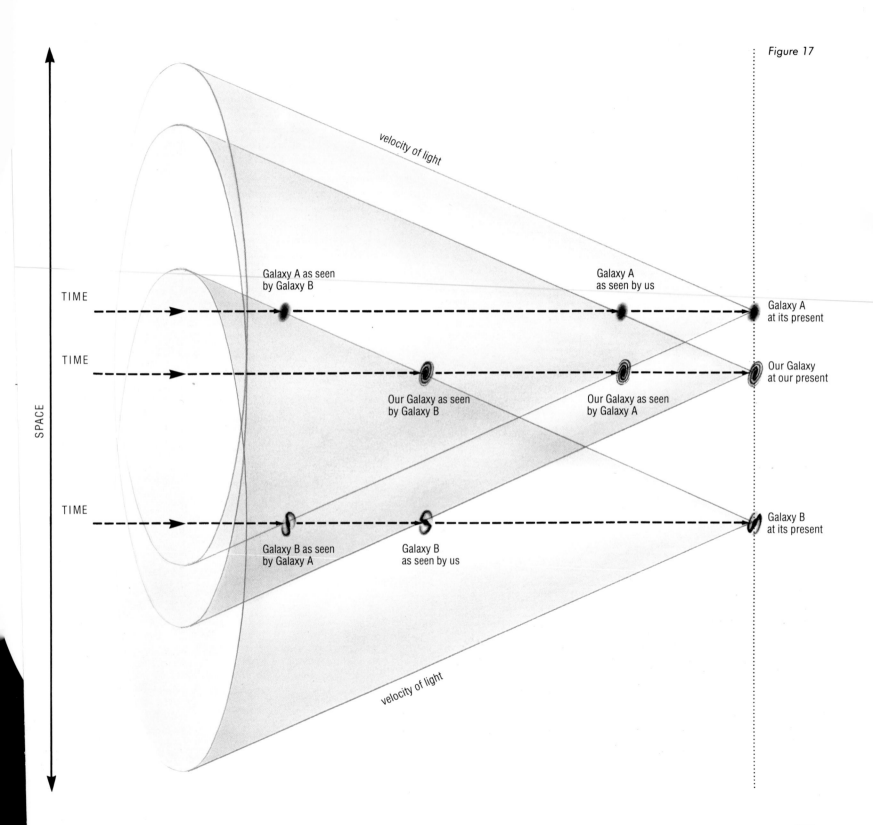

velocity of light

Galaxy A as seen
by Galaxy B

Galaxy A
as seen by us

Galaxy A
at its present

TIME

TIME

Our Galaxy as seen
by Galaxy B

Our Galaxy as seen
by Galaxy A

Our Galaxy
at our present

SPACE

TIME

Galaxy B as seen
by Galaxy A

Galaxy B
as seen by us

Galaxy B
at its present

velocity of light

Figure 17

Figure 19

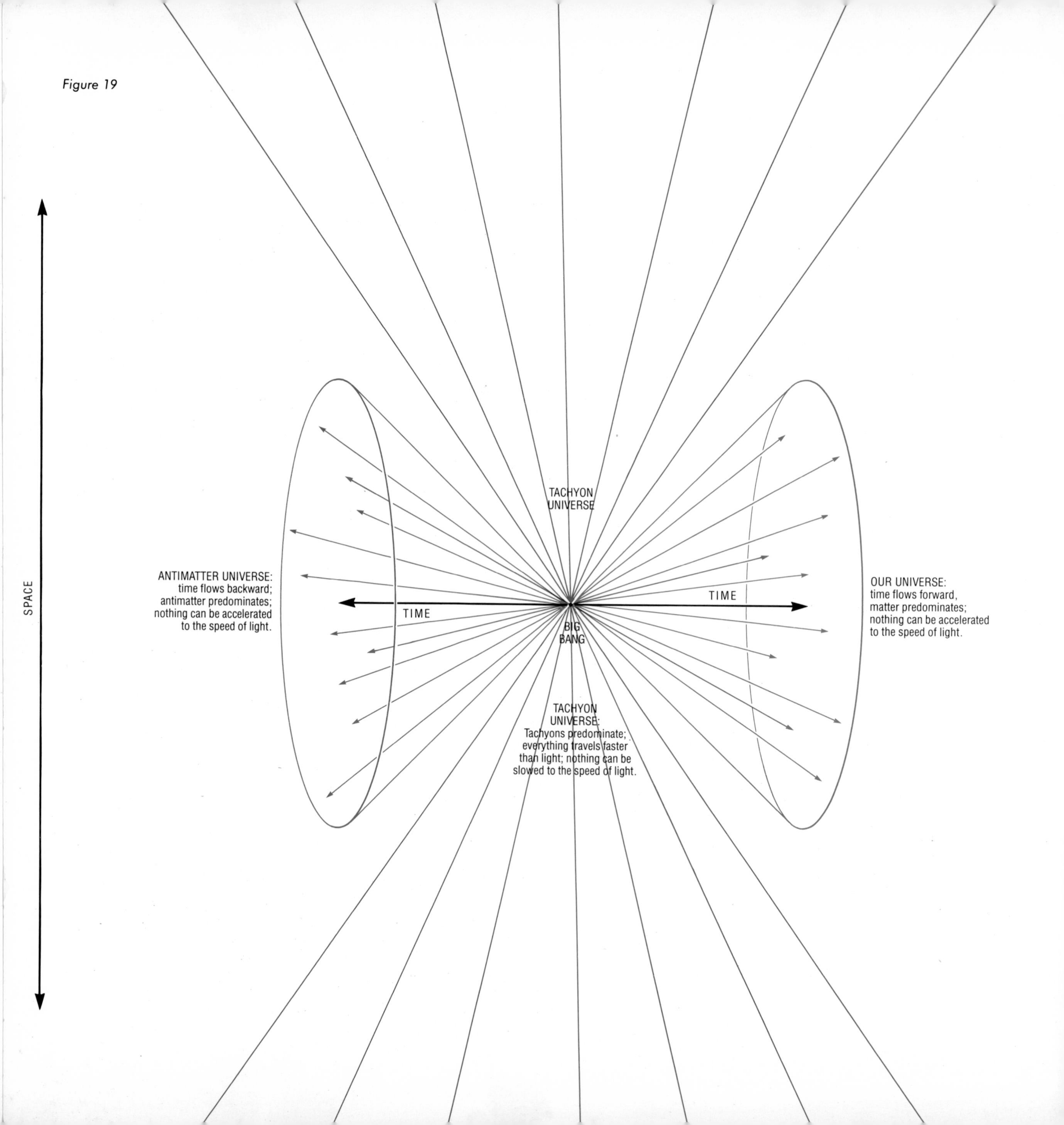

SPACE

TACHYON
UNIVERSE

ANTIMATTER UNIVERSE:
time flows backward;
antimatter predominates;
nothing can be accelerated
to the speed of light.

TIME

TIME

OUR UNIVERSE:
time flows forward;
matter predominates;
nothing can be accelerated
to the speed of light.

BIG
BANG

TACHYON
UNIVERSE:
Tachyons predominate;
everything travels faster
than light; nothing can be
slowed to the speed of light.

Modeling the Universe

A universe of questions about the universe remains. Evidence such as the quasar cutoff point and the cosmic background radiation indicate that those cosmologists are on the right track who maintain that the universe is finite in its age and material population, that its expansion began in an eruptive moment from which has unwound a process of cosmic evolution that continues today, and that the geometry of the universe may be more nearly understood by invoking orders of dimension beyond those that have served us locally here on earth. But within the boundaries of these assertions, even if they are quite correct, lies broad territory for cosmological investigation. Figure 19 illustrates just one of a multitude of imaginative cosmological models that may be constructed within the parameters of the big-bang hypothesis. The creation of J. Richard Gott III of Princeton University, this model proposes that not one but three universes sprang from the big bang. The Gott model attempts to account for two odd facts about our universe that trouble many cosmologists. One of these is that while the basic equations of physics are time-symmetrical—that is, they can be run forward or backward in time with equal efficacy—in the real universe time, it seems, moves in one direction only. The second oddity confronted by Gott is the scarcity of antimatter in our universe. For every sort of subatomic particle of matter, it is possible to conceive of a particle with the same mass but with opposite charge—an antiparticle. Yet only mere traces of antimatter have been found in nature. Why should nature be so nonsymmetrical, favoring matter over antimatter, running time in one direction but not the other?

Acting on a clue offered by theoretical physicists who maintain that antimatter can be thought of as ordinary matter moving in reverse time, Gott constructed his three-universe cosmology. He suggests that the big bang generated not only our universe, but also a second universe composed of antimatter and evolving in reverse time, as well as a third universe made up exclusively of particles that travel faster than light. The fleet particles of this ghostly third universe, called tachyons, are permissible under relativity theory, which requires only that nothing in our universe can be *accelerated* to the velocity of light; tachyons need not worry about this provision, for they have *always* been going faster than the speed of light. They occupy a mirror universe where everything travels faster than light and nothing can be reined to a velocity as slow as that of light. The Gott cosmology is a

masterpiece of symmetry without being dictatorial about it; it predicts, for instance, that there should be traces of contamination of our universe by antimatter (as has been verified by observation) and by tachyons (as has not). Whatever likelihood we might care to assign to its validity, in this combination of symmetry and imperfection the Gott model is redolent of nature's style.

The cosmological theories of today may be looked upon by our descendants with respect, bemusement, scorn or even hilarity, but the important thing is that cosmological endeavors need no longer be purely speculative in nature. We have learned how to test them against the real universe. And the universe is turning out to be remarkably amenable to such investigation. For all we knew until quite recently in human history, the cosmos at large might have been cramped or paltry, expressionless or opaque, unchanging or unpredictable. Instead, we find it lucid, intelligible, observable, evolving, and involving. The universe invites inquiry, as clear skies invite birds to fly.

Figure 19. A Three-Universe Big-Bang Cosmology
One of a number of sophisticated cosmological models that have been constructed within the broad purview of the big-bang account of cosmic history, this theory, proposed by J. Richard Gott III of Princeton University, postulates the existence of not one universe but three. It envisions that the big bang gave rise not only to our universe, where matter predominates over antimatter and time runs forward, but a second universe where antimatter predominates and time moves backward, as well as a tachyon universe where everything moves faster than the velocity of light. Our universe and the antimatter universe are segregated in terms of time. Both are segregated from the tachyon universe in terms of space, since the tachyons in the first instant of creation fled beyond the light cones of all observers in both the matter and antimatter universes.

143 (overleaf) Quasars are probably the nuclei of young galaxies caught in the act of shedding tremendous amounts of energy as they condense out of primordial gas. Here two quasars have been recorded at X-ray wavelengths by detectors aboard an earth satellite; the colors are not genuine, since X-rays lie well up the electromagnetic spectrum from light and so cannot be said to have color, but have been generated by computer. The dim quasar in the upper left corner is fifteen billion light-years away; its energy has been traveling through space for roughly three-quarters of the time since the expansion of the universe began.

What we have learned
Is like a handful of earth;
What we have yet to learn
Is like the whole world.

—AVVAIYAR

Sources of Photographs

The author wishes to thank the following for their generosity in making their photographs available for this book. All photographs are copy- right © the individuals or institutions named, and may not be repro- duced without their permission.

Photo Number	Object	Source
Frontispiece	Galaxy NGC6744	Raymond J. Talbot, Jr., Reginald J. Dufour, and Eric B. Jensen, Rice University
1	The Sun	United States Naval Research Laboratory
2	Milky Way	Mssrs. Brodkorb, Rihm, and Rusche, Astrophoto Laboratory
3	Coalsack Nebula	Harvard College Observatory
4	Cone Nebula	Hale Observatories
5	Horsehead Nebula	Royal Observatory, Edinburgh
6	Orion Nebula	Royal Observatory, Edinburgh
7	Trapezium	Lick Observatory
8	Eagle Nebula	Hale Observatories
9	Rosette Nebula	Hale Observatories
10	Rosette Nebula, Interior	Kitt Peak National Observatory
11	Eta Carina Nebula	Association of Universities for Research in Astronomy, Inc., Cerro Tololo Inter-American Observatory
12	Eta Carina Nebula, Central Portion	Association of Universities for Research in Astronomy, Inc., Cerro Tololo Inter-American Observatory
13	Trifid and Lagoon Nebulae	Hale Observatories
14	Lagoon Nebula	Association of Universities for Research in Astronomy, Inc., Kitt Peak National Observatory
15	Trifid Nebula	David Malin, Anglo-Australian Telescope Board
16	Star Cluster NGC3293	David Malin, Anglo-Australian Telescope Board
17	Pleiades Star Cluster	Hale Observatories
18	Globular Star Cluster	Mt. Stromlo and Siding Spring Observatories, Australian National Observatory
19	Globular Star Cluster M13	United States Naval Observatory
20	Globular Star Cluster M3	Hale Observatories
21	Globular Star Cluster M15	Kitt Peak National Observatory
22	Globular Star Cluster Omega Centauri	Cerro Tololo Inter-American Observatory
23	Globular Star Cluster M5	Kitt Peak National Observatory
24	Globular Star Cluster 47 Tucana	Cerro Tololo Inter-American Observatory
25	Globular Star Clusters NGC6522 and NGC6528	Kitt Peak National Observatory
26	Intergalactic Globular Star Cluster NGC2419	Rene Racine, Hale Observatories
27	"Planetary" Nebula M27	Hale Observatories
28	"Planetary" Nebula M57	Hale Observatories
29	Veil Nebula Field View	Hale Observatories
30	Veil Nebula Detail	Kitt Peak National Observatory
31	Crab Nebula Superimposition	Guido Münch and Walter Baade, Hale Observatories
32	Crab Nebula	Lick Observatory
33	Milky Way in Cygnus	National Geographic-Palomar Observatory Sky Survey
34	Milky Way in Sagittarius	National Geographic-Palomar Observatory Sky Survey
35	Milky Way Mosaic	Mt. Stromlo and Siding Springs Observatories, Australian National University
36	Large Magellanic Cloud	Raymond J. Talbot, Jr., Reginald J. Dufour, and Eric B. Jensen, Rice University

Photo Number	Object	Source
37	The Magellanic Clouds	Harvard College Observatory
38	Small Magellanic Cloud	Royal Observatory, Edinburgh
39	Small Magellanic Cloud	Raymond J. Talbot, Jr., Reginald J. Dufour, and Eric B. Jensen, Rice University
40	Galaxy M31	Hale Observatories
41	M31 Central Regions	Association of Universities for Research in Astronomy, Inc., Kitt Peak National Observatory
42	Galaxy M31 Central Region	Hale Observatories
43	M31 Near Nucleus	Hale Observatories
44	M31 Outer Arms	Hale Observatories
45	M31 Nucleus	Lick Observatory
46	M32	Kitt Peak National Observatory
47	NGC205	Hale Observatories
48	NGC147	Hale Observatories
49	NGC185	Lick Observatory
50	M31 Radio Map	Elly M. Berkhuijsen, Max-Planck-Institut für Radioastronomie
51	M33	Hale Observatories
52	Sculptor Dwarf Spheroidal Galaxy	European Southern Observatory
53	M101	Hale Observatories
54	NGC7331	Hale Observatories
55-62	Galaxies by Formal Type	Hale Observatories
63	NGC2841	Hale Observatories
64	NGC2613	Allan Sandage, Hale Observatories
65	M64	Hale Observatories
66	NGC3623	United States Naval Observatory
67	M104	Steven Strom, Kitt Peak National Observatory
68	NGC4565	Lick Observatory
69	NGC3992	Lick Observatory
70	NGC4541	Steven Strom, Kitt Peak National Observatory
71	NGC1360	European Southern Observatory
72	NGC4650	Laird A. Thompson, Kitt Peak National Observatory
73	NGC4548	Laird A. Thompson, Kitt Peak National Observatory
74	M83	Raymond J. Talbot, Jr., Reginald J. Dufour, and Eric B. Jensen, Rice University
75	M83 Negative Print	Royal Observatory, Edinburgh
76	M83 Radio Map	D.H. Rogstad, California Institute of Technology
77	NGC4477	Steven Strom, Kitt Peak National Observatory
78	M84	Kitt Peak National Observatory
79	M49	Kitt Peak National Observatory
80	Carina Dwarf Galaxy	Royal Observatory, Edinburgh
81	NGC3077	Kitt Peak National Observatory
82	Sextans Dwarf Galaxy	Hale Observatories
83	NGC5364	Kitt Peak National Observatory
84	NGC6744	Cerro Tololo Inter-American Observatory
85	M74	Hale Observatories
86	NGC2683	United States Naval Observatory
87	NGC5907	United States Naval Observatory
88	Pluto & NGC5248	K. Alexander Brownlee
89,90	NGC4725 with Supernova	Hale Observatories
91	NGC4096 with Supernova	Lick Observatory
92	NGC4303 with Supernova	Lick Observatory
93	Centaurus A	Hale Observatories
94	Centaurus A	Raymond J. Talbot, Jr., Reginald J. Dufour, and Eric B. Jensen, Rice University
95	Perseus A	Hale Observatories
96	Cygnus A	Hale Observatories
97	M87 (large)	Malcolm Smith and W.E. Harris, Cerro Tololo Inter-American Observatory
98	M87	James Wray, McDonald Observatory, University of Texas
99	M87 (small)	Lick Observatory
100	M77	Lick Observatory
101	NGC1566	Harvard College Observatory
102	M106	James Wray, McDonald Observatory, University of Texas
103	M94	Kitt Peak National Observatory
104	NGC4151	Hale Observatories

Selected Bibliography

BOOKS ON GALAXIES AND RELATED SUBJECTS

Of General Interest

Abbott, Edwin, A., *Flatland*, New York, Dover Publications, 1952.

Allen, Richard Hinckley, *Star Names: Their Lore and Meaning*. New York, Dover Publications, 1963.

Berendzen, Richard Hart, and Daniel Seely, *Man Discovers the Galaxies*. New York, Science History Publications, 1976.

Bok, Bart J., and Priscilla F. Bok, *The Milky Way*, Cambridge, Mass., Harvard University Press, 1974.

Bondi, Hermann, *Relativity and Common Sense*. Garden City, N.Y., Anchor Doubleday, 1964.

Ferris, Timothy, *The Red Limit: The Search for the Edge of the Universe*. New York, William Morrow and Co., 1977.

Golden, Frederic, *Quasars, Pulsars, and Black Holes*. New York, Pocket Books, 1977.

Hoyle, Fred, *Galaxies, Nuclei, and Quasars*. New York, Harper and Row Publishers, 1965.

Maffei, Paolo, *Beyond the Moon*. Cambridge, Mass., MIT Press, 1978.

Mitton, Simon, ed., *The Cambridge Encyclopedia of Astronomy*. New York, Crown Publishers Inc., 1977.

Moore, Patrick, *The Amateur Astronomer*. New York, W.W. Norton and Company, 1968.

Page, Thornton, and Lou Williams Page, *Beyond the Milky Way: Galaxies, Quasars, and the New Cosmology*. New York, The Macmillan Company, 1969.

Page, Thornton, and Lou Williams Page, eds., *Stars and Clouds of the Milky Way: The Structure and Motion of Our Galaxy*. New York, The Macmillan Company, 1968.

Russell, Bertrand, *The ABC of Relativity*. New York, New America Library, 1970.

Scientific American, *New Frontiers in Astronomy*. San Francisco. W.H. Freeman and Company, 1975.

Shapley, Harlow, *Galaxies*. Cambridge, Mass., Harvard University Press, 1972.

Sullivan, Walter, *Black Holes: The Edge of Space, The End of Time*. Garden City, N.Y., Anchor Press/Doubleday, 1979.

Weinberg, Steven, *The First Three Minutes*. New York, Basic Books Inc. Publishers, 1977.

Whitney, Charles A., *The Discovery of Our Galaxy*. New York, Alfred A. Knopf, 1971.

Technical and Semi-Technical

Abetti, Giorgio, and M. Hack, *Nebulae and Galaxies*. New York, Thomas Y. Crowell Co., 1964.

Baade, Walter, *Evolution of Stars and Galaxies*. Cambridge, Mass., MIT Press, 1975.

Berkhuijsen, Elly M., and Richard Wielebinski, *Structure and Properties of Nearby Galaxies* (IAU Symposium no. 77). Boston, D. Reidel Publishing Company, 1978.

Clark, David H., and F. Richard Stephenson, eds. *The Historical Supernovae*. Oxford, Pergamon Press, 1977.

Dickens, R.J., and Joan E. Perry, eds., *The Galaxy and the Local Group* (Royal Greenwich Observatory Bulletin no. 182). Herstmonceux, Royal Greenwich Observatory, 1976.

Einstein, Albert, *The Meaning of Relativity*. Princeton, N.J., Princeton University Press, 1956.

Einstein, Albert, *Relativity: The Special and General Theory*. New York, Crown Publishers Inc., 1961.

Hazard, C., and S. Mitton, *Active Galactic Nuclei*. Cambridge, Cambridge University Press. 1979.

Hodge, Paul W., *The Physics and Astronomy of Galaxies and Cosmology*. New York, McGraw-Hill Book Company, 1966.

Lang, Kenneth R., and Owen Gingerich, eds., *A Sourcebook in Astronomy and Astrophysics, 1900-1975*. Cambridge, Mass., Harvard University Press, 1979.

Longair, M.S., and J. Einasto, *The Large Scale Structure of the Universe* (IAU Symposium no. 79). Boston, D. Reidel Publishing Company, 1978.

Middlehurst, Barbara M., and Lawrence H. Aller, *Nebulae and Interstellar Matter* (volume 7 of *Stars and Stellar Systems*). Chicago, University of Chicago Press, 1968.

Mitton, Simon, *Exploring the Galaxies*. New York, Charles Scribner's Sons, 1976.

North, J.D., *The Measure of the Universe: A History of Modern Cosmology*. Oxford, Oxford University Press, 1955.

O'Connell, D.J.K., ed., *Study Week on Nuclei of Galaxies*. Amsterdam, North-Holland Publishing Co., 1971.

Payne-Gaposchkin, Cecilia, *Stars and Clusters*. Cambridge, Mass., Harvard University Press. 1979.

Sandage, Allan, Mary Sandage, and Jerome Kristian, eds., *Galaxies and the Universe* (volume 9 of *Stars and Stellar Systems*). Chicago, University of Chicago Press, 1975.

Setti, Giancarlo, ed., *Structure and Evolution of Galaxies*. Boston, D. Reidel Publishing Company, 1975.

Shakescraft, John, ed., *The Formation and Dynamics of Galaxies* (IAU Symposium no. 58). Boston, D. Reidel Publishing Company, 1974.

Shapley, Harlow, *The Inner Metagalaxy*. New Haven, Yale University Press, 1957.

Shapley, Harlow, ed., *Sourcebook in Astronomy, 1900-1950*. Cambridge, Mass., Harvard University Press, 1960.

Shklovskii, Iosif S., *Stars: Their Birth, Life and Death*. San Francisco, W.H. Freeman & Co., 1978.

Shklovskii, Iosif S., *Supernovae*. New York, John Wiley and Sons, 1968.

Tayler, R.J., *Galaxies: Structure and Evolution*. New York, Crane, Russak and Company, 1978.

Tinsley, Beatrice M., and Richard B. Larson, eds., *The Evolution of Galaxies and Stellar Populations*. New Haven, Yale University Observatory, 1977.

Unsöld, Albrecht, *The New Cosmos*. New York, Springer-Verlag, 1977.

Woltjer, Lodewijk, ed., *Galaxies and the Universe*. New York, Columbia University Press, 1968.

PERIODICALS

Of General Interest

Astronomy. Milwaukee, Wisconsin, AstroMedia Corp.
Cosmic Search. Delaware, Ohio, Cosmic-Quest Inc.
Mercury. San Francisco, Astronomical Society of the Pacific.
Scientific American. New York, Scientific American Inc.
Sky & Telescope. Cambridge, Mass., Sky Publishing Corp.
Spaceflight. London, The British Interplanetary Society.

Technical

Annual Review of Astronomy and Astrophysics. Palo Alto, Calif., Annual
  Reviews Inc.
Astronomical Journal. New York, American Institute of Physics.
The Astrophysical Journal. Chicago, University of Chicago Press.
Journal of the Royal Astronomical Society. Oxford, Blackwell Scientific
  Publications.
Journal of the Royal Astronomical Society of Canada. Toronto, Royal
  Astronomical Society.
Monthly Notices of the Royal Astronomical Society. Oxford, Blackwell
  Scientific Publications.
Publications of the Astronomical Society of the Pacific. San Francisco,
  Astronomical Society of the Pacific.
Soviet Astronomy. New York, American Institute of Physics.
Vistas in Astronomy, New York, Pergamon Press.

TEXTBOOKS

Abell, George, *Exploration of the Universe,* 3rd ed. New York, Holt,
  Rinehart and Winston, 1975.
Field, George B., Gerrit L. Verschuur, and Cyril Ponnamperuma, *Cosmic
  Evolution: An Introduction to Astronomy.* Boston, Houghton Mifflin
  Co., 1978.
Hartmann, William K., *Astronomy: The Cosmic Journey.* Belmont, Calif.,
  Wadsworth Publishing Company, 1978.
Motz, Lloyd, and Anneta Duveen, *Essentials of Astronomy,* 2d ed. New
  York, Columbia University Press, 1977.
Roy, A. E., and D. Clarke, *Astronomy: Structure of the Universe.* New
  York, Crane, Russak and Company, 1977.

ATLASES AND CATALOGUES

Of General Interest

Becvar, Antonin, *Atlas of the Heavens.* Cambridge, Mass., Sky Publishing
  Corp., 1962
Becvar, Antonin, *Atlas of the Heavens—II. Catalogue.* Cambridge, Mass.,
  Sky Publishing Corp., 1964.
Norton, Arthur P., and J. Gall Inglis, *Norton's Star Atlas and Reference
  Handbook.* Cambridge, Mass., Sky Publishing Corp., 1966.
Rey, H. A., *The Stars: A New Way to See Them,* 3d ed. Boston, Houghton
  Mifflin, 1967.

Sandage, Allan, *The Hubble Atlas of Galaxies.* Washington, D.C.,
  Carnegie Institution of Washington, 1961.

Technical and Semi-Technical

Arp, Halton, *Atlas of Peculiar Galaxies.* Pasadena, California Institute
  of Technology, 1978.
Salentic, Jack W., and William G. Tifft, *The Revised New General
  Catalogue of Nonstellar Astronomical Objects.* Tucson, University
  of Arizona Press, 1973.
de Vaucouleurs, Gerard, and Antoinette de Vaucouleurs, *Reference
  Catalogue of Bright Galaxies.* Austin, University of Texas Press,
  1964.
de Vaucouleurs, Gerard, Antoinette de Vaucouleurs, and Harold G.
  Corwin, Jr., *Second Reference Catalogue of Bright Galaxies.*
  Austin, University of Texas Press, 1976.
Zwicky, F., E. Herzog, and P. Wild, *Catalogue of Galaxies and Clusters
  of Galaxies.* Pasadena, Caltech University Press, 1960-1968.

EPIGRAMMATICAL MATERIAL

Blyth, R. H., *Zen and Zen Classics.* Tokyo, Hokuseido Press, 1964.
Horace, *Satires, Epistles, and Ars Poetica.* Cambridge, Mass., Harvard
  University Press, 1978.
Jonson, Ben, "The Vision of Delight," in W. H. Auden and Normal
  Holmes Pearson, eds., *Medieval and Renaissance Poets.* New
  York, Penguin Books, 1978.
Lao Tzu, *Tao Te Ching,* Gia-Fu Feng and Jane English, translators. New
  York, Alfred A. Knoft, 1974.
Lucretius, *De Rerum Natura,* Cyril Bailey, translator. Oxford, Clarendon
  Press, 1972.
Sagan, Carl, Frank Drake, Ann Druyan, Timothy Ferris, Jon Lomberg,
  and Linda Salzman Saga, *Murmurs of Earth: The Voyager
  Interstellar Record.* New York, Random House, 1978.
Shakespeare, William, "Romeo and Juliet," in *The Complete Works.*
  Baltimore, Penguin Books, 1969.
Waley, Arthur, *The Way and Its Power: A Study of the Tao Te Ching and
  Its Place in Chinese Thought.* London, George Allen and Unwin,
  1968.
Whitney, Charles A., *The Discovery of Our Galaxy.* New York, Alfred
  A. Knopf, 1971.
Wilder, Thornton, *Our Town,* New York, Harper and Row, 1960.

PHOTOGRAPHS

Prints, slides, transparencies and in some cases posters of galaxies and
other astronomical objects may be ordered from the following sources:

European Southern Observatory
c/o CERN
Geneva, Switzerland

Hale Observatories
Pasadena, California

Kitt Peak National Observatory
Tucson, Arizona

Lick Observatory
University of California
Santa Cruz, California

United States Naval Observatory
Washington, D.C.

Glossary

It's not easy to describe
the sea with the mouth
—Kokyū

Astronomical terminology can be confusing for those who are not trained in the field and far from perfectly lucid for those who are. Sometimes one word is applied to more than one sort of object; astronomers speak of interstellar "clouds" composed of gas and dust, but the Magellanic Clouds are galaxies and the term "cloud" may also mean a cluster of galaxies. Sometimes several names are applied to a single object, as when open star clusters are referred to as "galactic" clusters, or when the Local Group of galaxies is termed not a group but a cluster. Alternately, a single object may be bedecked with many designations. For example, the enormous galaxy M87 owes its name to the eighteenth-century catalogue of Charles Messier, but it is also known as NGC4486, after the New General Catalogue of 1888, 3C274 for the Third Cambridge Catalogue of Radio Sources, and as Virgo A, signifying that it is the most powerful source of radio energy in the constellation Virgo.

Misnomers abound. Many result from mistaken first impressions, like those that produced Columbus's "Indians." "Quasars" were so named because at first blush they look "quasi-stellar." By the time it became clear that they were probably not stars but rather the nuclei of distant galaxies the name had taken hold and it was too late to change it. The "planetary" nebulae are envelopes of gas disgorged by aging stars; first impressions to the contrary, they are about as unplanetary as anything can be.

I have tried to keep technical terminology to a minimum, and in one case I have knowingly suppressed important information in the interest of painting a simple picture: In the "journey" sections many of the perceptual distortions produced by relativistic space flight have been ignored, such as the blue-shifting of galaxies ahead and the red-shifting of those astern, in order to concentrate attention on the galaxies rather than on the effects of space flight. Technically-inclined readers are asked to forgive this omission.

Those impatient with technical terminology are invited to employ this glossary to cope with what little of it has proved unavoidable, and to take comfort in the reflection that the designations of galaxies, like the Latin names of plants, are but human inventions and remain unknown to that which they designate.

A. Denotes the most powerful source of radio energy within a given constellation, as in Cygnus A or Centaurus A.

Arp. Designates galaxies listed in the *Atlas of Peculiar Galaxies*, by the astronomer Halton Arp.

Atom. The smallest unit of matter of a chemical element. Atoms may be broken down into subatomic particles, but once this happens they will lose the chemical properties characteristic of their element. The repertoire of possible chemical interactions is vastly enhanced when atoms are combined into molecules. Many sorts of molecules, as well as free atoms, are found floating in space.

Barred Spiral. A spiral galaxy characterized by a prominent realm of stars and interstellar material arranged in the shape of a bar or spindle projecting out from either side of the central bulge. The bar is probably created by dynamical interactions in the overall gravitational environment of the galaxy. Most spiral galaxies show at least a trace of a bar. The term "barred" is reserved for those in which this feature is prominent.

Billion. The billion employed in this book is the American billion, equal to one thousand million, or 10^9.

Binary galaxies. A pair of galaxies bound together gravitationally. Normally they will coexist peacefully in orbit around their common center of gravity, but occasionally they will pass close to each other, producing spectacular distortions in their structures. The Milky Way and Andromeda galaxies form a binary pair.

Black hole. An object compressed to so high a density as to imprison even its own light. A black hole comes into existence when a collapsing star or other object wraps itself in a gravitational field sufficiently intense that the velocity required to escape from it exceeds the velocity of light, so that nothing can escape. Although the term invokes romantic imaginings of "holes in space," a black hole is at its heart quite a substantial object.

Bright nebula. See *Nebula*.

C. Designates radio sources listed in one of several Cambridge Catalogues of Radio Objects. Successive catalogues are designated by numbers preceding the letter designation, so that 3C273 means that the object —in this case a quasar—is number 273 in the Third Cambridge Catalogue.

Central bulge. The elliptical region at the center of a spiral galaxy, situated something like the yolk of a fried egg. Also called the lens.

Cepheid variable star. A supergiant pulsating star that varies in brightness. There are several classes of Cepheids, each valuable to astronomers in that the amount of time it takes them to go through a cycle of brightness variation is directly related to their intrinsic brightness. An astronomer seeking to determine the distance of a nearby galaxy can measure the period of variability of Cepheids there, derive their true brightness, compare this with their apparent brightness in the sky —the rule is that the apparent brightness of a star diminishes with the square of its distance—and so determine the distance of the star and of its galaxy. Cepheids are bright enough to be identified with existing telescopes in galaxies at distances of up to about ten million light-years. Polaris, the North Star, is a Cepheid variable.

Cloud. Alternate term for a cluster of galaxies. Also used informally for interstellar material within a galaxy, as in the "Monoceros Dark Cloud." The Magellanic Clouds are galaxies.

Cluster. An association of many galaxies bound together gravitationally. The gravitational bonds of a cluster of galaxies are strong enough to sustain the association against the expansion of the universe, so that the expansion takes place not within the cluster but in the spaces between clusters.

Constellation. A configuration of stars in the sky usually recognized as tracing out a recognizable figure or symbol. For convenience, modern star charts divide the entire sky into constellations. But constellations have little astrophysical significance since they tell us only where stars appear in the sky, not where they are relative to one another in real space. For example, the stars of Orion's belt are about sixteen hundred light-years away, while the distance to the star Betelgeuse at Orion's right shoulder is only five hundred twenty light-years, and Rigel, his left foot, is nine hundred light-years away.

Corona. See *Halo.*

Cosmic rays. Charged subatomic particles—most are protons—streaking through space at velocities approaching that of light. Eruptions on the surface of the sun are known to produce cosmic rays, as are supernovae, but other as yet unidentified sources are thought to exist.

Cosmology. The study of the structure and history of the universe at large. It can be subdivided into theoretical cosmology, which looks at the mathematical and physical possibilities of how the universe might be constructed, and observational cosmology, which gathers astronomical data relevant to cosmological questions. In practice, contributions to cosmology come from astronomers, astrophysicists, mathematicians and theorists working in a variety of fields and with a variety of styles.

Cosmos. A term for the universe that emphasizes the belief in an orderly underlying structure to all of creation. It comes from the Greek *Kósmos,* meaning 'order.'

Dark nebula. See *Nebula.*

Degenerate star. A star that has used up most of its nuclear fuel and has collapsed to a state of high density.

Density wave. In spiral galaxies, a wave propagated through the interstellar material of the disk in a spiral pattern. The wave promotes the collapse of interstellar clouds into new stars; the stars in turn light up the surrounding interstellar medium, creating the visible phenomenon we call the spiral arms. The density wave is thought to be generated by resonances in the gravitational interaction of the stars of the galaxy as they move in their orbits.

Diffuse nebula. See *Nebula.*

Disk. The flattened component of a spiral galaxy, home to billions of stars and to large tracts of interstellar material. See *Galaxy, spiral.*

Doppler shift. Displacement in the apparent wavelength of light or other radiation coming from a body that is in relative motion toward or away from the observer. If the object is approaching, its light will be compressed and will appear shorter in wavelength than if it were at rest. If it is receding, the opposite effect occurs and the light is shifted toward the long-wavelength, or red, end of the spectrum. Red shifts in the light of distant galaxies provide evidence that the universe is expanding.

Dwarf galaxy. See *Galaxy, dwarf.*

Dwarf star. This deceptively diminutive term is employed by astrophysicists to apply to most normal stars like our sun. In general usage, the term is often encountered with a modifier, as in white dwarf or black dwarf, which are degenerate stars that have collapsed to a size comparable to that of the earth.

Electron. A negatively charged subatomic particle that when found in an atom orbits the nucleus.

Electromagnetic spectrum. See *Spectrum.*

Elliptical galaxy. See *Galaxy, elliptical.*

Equator, galactic. The plane of the disk of the Milky Way Galaxy. The earth's equator is inclined sixty-three degrees relative to the galactic equator.

Event horizon. The boundary around a black hole from within which no matter or information can escape.

Evolution, stellar. The development of a star from its origin as a protostar, or recently collapsed ball of gas, through its career until it runs out of hydrogen and helium fuel and ebbs into darkness. For a normal star like our sun, this process takes billions of years, most of it spent on what astrophysicists call the "main sequence" where the star maintains a stable balance between the gravitational and radiative forces within it. When its fuel is exhausted, a star with the mass of the sun leaves the main sequence and expands enormously to become a "red giant" star. Then it ventilates much of its atmosphere into space and what remains of it settles down into a "white dwarf" star. The term "evolution" has been criticized by some researchers who point out that stars are not subject to Darwinian selection, but it remains useful in discussing stellar processes at large, if not the development of individual stars.

Fission, nuclear. See *Nucleus, atomic.*

Flare star. Dim dwarf stars that produce sudden, irregular outbursts of energy. They are probably stars that have recently formed and are still subject to imbalances between the gravitational force that tends to collapse them and the radiative pressure that tends to sustain them against collapse.

Fusion, nuclear. See *Nucleus, atomic.*

Galactic nucleus. See *Nucleus, galactic.*

Galactic star cluster. See *Star cluster, galactic.*

Galaxy. A giant association of stars and interstellar gas and dust. In mass, galaxies range from roughly ten million to perhaps ten thousand billion times that of the sun.

Galaxy, dwarf. A small, dim galaxy. Difficult to define with exactitude, dwarf galaxies are smaller than major galaxies like the Milky Way but larger than globular star clusters.

Galaxy, elliptical. A galaxy whose stars are arranged in an elliptical volume of space. Unlike the flattened spirals, ellipticals have no disk, no spiral arms and relatively little interstellar material. In shape they range from nearly spherical to almost cigar-shape.

Galaxy, irregular. A disorderly looking galaxy that displays little of the symmetry of ellipticals or spirals. Most irregulars are dwarves. Often they are satellites of larger galaxies.

Galaxy, spiral. A galaxy possessing a flattened disk marked by a pattern of spiral arms. In addition to stars, the disk contains interstellar clouds of gas and dust. The spiral arms are luminous areas within the interstellar medium where the clouds have been compressed sufficiently to trigger the foundation of stars, whose light in turn traces out the pattern of the arms.

Gamma rays. The highest-energy form of electromagnetic radiation, extremely high in frequency, short in wavelength. See *Spectrum.*

Gaseous nebula. See *Nebula.*

Globular star cluster. See *Star cluster, globular.*

Globule. A dark ball of interstellar dust and gas, often found in the vicinity of star-forming nebulae. In many cases, globules appear to be collapsing on their way to forming new stars. They have been described as dust balls rolled up in the turbulence of a collapsing cloud.

Gravitation. The universal attraction of particles of matter for one another. See *Gravity* and *Relativity, theories of.*

Gravity. The universal attraction of matter for matter. Like light and other radiation, the force of gravity decreases by the square of the distance, so that if the distance separating two galaxies is doubled, their gravitational attraction to each other will be reduced to one-quarter its original value.

Group.   A small cluster of galaxies.

Halo.   A spheroidal zone surrounding a galaxy and inhabited by old stars, globular clusters and clouds of gas. Also called the corona.

Helium.   After hydrogen, the second simplest and second most abundant element in the universe.

H II region.   A bright cloud of predominantly hydrogen gas that glows by virtue of its atoms having absorbed energy from nearby stars and re-emitted it, as occurs in a neon light. Normally these are regions where recently formed stars are pouring energy into the surrounding cloud from which they were created. In this book, H II regions are referred to by the broader term "bright nebulae." See *Nebula*, also *Hydrogen*.

Hubble constant.   A measure of the rate of expansion of the universe. Modern estimates cite the Hubble constant at 50 kilometers per second per megaparsec. This means that for every megaparsec (i.e. 3.26 million light-years) farther out one looks, one finds galaxies receding at an additional 50 kilometers (31 miles) per second.

Hubble's law.   The rule that the light from distant galaxies is red-shifted to a degree proportionate to their distance from us. Discovered by Edwin Hubble in 1929, this was the first indication of the expansion of the universe. The Hubble law is not valid within clusters of galaxies, which are gravitationally bound together and proof against the expansion of the universe, but comes into play in the realm of "pure Hubble flow" between superclusters of galaxies.

Hydrogen.   The simplest and least massive of atoms, normally consisting of one proton and one electron. Hydrogen is by far the most common element in the universe. When a cloud of hydrogen gas has been ionized—that is, when many of its atoms have gained or lost electrons, as they will when energized by the radiation of a nearby star—it is referred to in astronomy as an H II region, after the chemical symbol for ionized hydrogen. Bright nebulae like the Orion Nebula in the Milky Way Galaxy and the Tarantula Nebula in the Large Magellanic Cloud are H II regions.

IC.   Designates objects listed in the Index Catalogue, a supplement to the *New General Catalogue.*

Infrared light.   Electronic radiation lying just to the low-frequency side of visible light on the electromagnetic spectrum; heat. Young stars still wrapped in the clouds from which they formed can sometimes be observed at infrared wavelengths. See *Spectrum.*

Interacting galaxies.   Two or more galaxies that have drifted close enough together that their gravitational interaction has manifested itself in an obvious fashion, such as by structural distortions in each system or the exchange or expulsion of stars from them.

Interstellar medium.   Matter found in the spaces between stars. In a normal spiral galaxy like ours, the interstellar medium is composed primarily of hydrogen and helium gas, traces of more complicated atoms and molecules, and dust contributed by the explosions of dying stars.

Irregular galaxy.   See *Galaxy, irregular.*

Island universe.   A galaxy. Though wonderfully descriptive of both the dimensions and independent stature of galaxies, the term has fallen from use in favor of others more concise and less presupposing.

Latitude, galactic.   Coordinates specified in terms of degrees above or below the plane of the Milky Way Galaxy.

Lens.   See *Central bulge.*

Lenticular galaxy.   See *SO Galaxy.*

Light-year.   The distance traveled by light in one year. The velocity of light is 186,000 miles per second in a vacuum, and a light-year equals some 5.8×10^{12}, or nearly six thousand billion, miles.

Lookback time.   A term employed to call attention to the fact that we see remote astronomical objects as they were when their light left them long ago. A galaxy one hundred million light-years away appears to us as it was one hundred million years in the past, while a quasar at ten billion light-years lookback time is seen as it was ten billion years ago, when the universe was perhaps one half its present age.

Longitude, galactic.   Coordinates specified in terms of degrees along the plane of the galaxy measured eastward from the galactic center in the constellation Sagittarius.

M.   Designates objects listed in the catalogue of Charles Messier, originally published in 1781. A comet-hunter, Messier listed in his catalogue any fuzzy-looking object that might be mistaken for a comet. As a result, the catalogue contains a polyglot mixture of bright and dark nebulae, open and globular star clusters, planetary nebulae and galaxies.

Magnitude.   The brightness of a star or other astronomical object denoted on a logarithmic scale. A difference of five magnitudes equals a difference of 10^2 times in luminosity, while a difference of one magnitude equals a discrepancy in luminosity of 2.5 times. Objects brighter than zero magnitude are designated by minus numbers. The apparent magnitude of Sirius, the brightest star in the skies of earth after the sun, is minus 1.6; the North Star, Polaris, has a magnitude of 2; the dimmest stars visible to the unaided eye are about sixth magnitude. Large telescopes can detect objects at twenty-fourth magnitude and even dimmer.

Magnitude, absolute.   The magnitude of a star or other astronomical body as it would appear if viewed at a standard distance of ten parsecs or 32.6 light-years. See *Magnitude.*

Magnitude, apparent.   The magnitude of a star as it appears in the sky. See *Magnitude.*

Mass.   The total amount of matter in a body. In this book, the masses of galaxies often are expressed in terms of their population of stars, for instance by stating that a galaxy has approximately one hundred billion stars like the sun. This assumes that the sun is a star of typical mass, a not wildly unlikely assumption. But if galaxies typically have a great number of low-mass dwarf stars too dim as yet to have been observed, as some astronomers suspect, then a galaxy with a mass one hundred billion times that of the sun would actually have many more than one hundred billion stars.

Microwaves.   See *Spectrum.*

Milky Way.   Our view of the disk of our galaxy, a softly glowing band of starlight stretching across the sky. By extension, our galaxy as a whole.

Nebula.   Originally, any patch of hazy, diffuse light in the sky. A number of quite different objects fall under the umbrella of the term. Some are clouds of gas which have been excited to glow by hot stars within them. These are referred to as "bright nebulae" in this book. Astrophysicists designate them H II regions, after the chemical symbol for ionized hydrogen. Other bright nebulae consist of gas ejected from dying stars as "planetary" nebulae or, in the case of more violent stellar explosions, as supernovae remnants. H II regions, supernovae remnants and "planetary" nebulae are lumped together by astronomers under the term "gaseous nebulae." Interstellar clouds that glow not by emitting their own light but by reflected starlight are called "reflection" or "diffuse" nebulae. Clouds that do not glow at all are called "dark nebulae." Adding to the burden upon this single word is the fact that galaxies, in the days before telescopes could be built that were capable of resolving them into stars, also were classified as nebulae; even today one encounters the anachronistic term "spiral nebulae" for a spiral galaxy.

Neutron star. A degenerate star that has collapsed to extremely high density. A neutron star with the mass of the sun would measure only about twelve miles in diameter. Pulsars are believed to be neutron stars emitting energy at radio and other wavelengths as they spin; the energy spirals outward through the magnetic field of the pulsar like the jet of a water sprinkler, showing up to an outside observer as a pulse each time the streamer flashes past.

NGC. Designates objects listed in the New General Catalogue of Nonstellar Astronomical Objects.

Nova. An explosion of a star powerful enough that it dramatically if briefly increases the brightness of the star but mild enough so that it leaves a working star behind afterward. Novae were described by the late Cecelia Payne-Gaposchkin of Harvard University as "probably very old stars that are taking a drastic way out from an intolerable state, when they can no longer support themselves in the style to which they have been accustomed."

Nucleus, atomic. The center of an atom, around which whirl clouds of electrons. Considerable amounts of energy are employed in binding together the particles of the nucleus. Fission reactors and the atomic bomb work by breaking down the nucleus and releasing some of this energy; hydrogen bombs and stars release still more energy by fusing nuclei together.

Nucleus, galactic. The center of a galaxy. Galactic nuclei are typically small and bright. Their nature is not yet well known. Hypotheses concerning the anatomy of galactic nuclei range from dense star clusters to black holes.

Open star cluster. See *Star cluster, open.*

Parallax. The angle described by one astronomical unit—the mean distance from the earth to the sun—as viewed from a nearby star. Astronomers measure interstellar distances by photographing neighboring stars from alternate sides of the earth's orbit and measuring the displacement that this change of perspective introduces in their apparent location against the background of more distant stars.

Parsec. A unit of distance equal to 3.26 light-years. An abbreviation for parallax-second, a parsec is equal to the distance at which one astronomical unit—the mean distance between the earth and sun—describes an angle of one second of arc. Astronomers normally, for convenience, employ the parsec rather than the light-year as a unit of distance, and some argue that it would be simpler to expunge the latter term from the astronomical vocabulary. But both quantities are based upon the earth's orbit, arbitrary by transstellar standards.

Peculiar galaxy. A galaxy that fails to fit into one of the structural categories of spiral, elliptical, SO or irregular.

Planet. A body orbiting a star and shining by its reflected light. Planets up to about fifty times the mass of Jupiter might exist; in objects more massive than that, nuclear processes would begin at the core and they would become stars. A lower limit to the mass that a body must have to qualify as a planet has not yet been established, since the question has not come up here in the solar system and it is not yet possible to observe planets of other stars. In the solar system, the term "minor planet" or "asteroid" is applied to thousands of lesser bodies orbiting the sun, all of them much smaller than even the smallest planet, Pluto. Comets are objects of still lower density, most of which orbit the sun at great distances.

"Planetary" nebula. A shell of gas ejected into space by a star that has consumed much of its hydrogen fuel and has lost its internal balance. See *Nebula.*

Poles, galactic. The axis of the Milky Way Galaxy defined by drawing an imaginary line through the galactic nucleus perpendicular to the plane.

Proton. A heavy subatomic particle with a positive charge, found in the nuclei of atoms.

Pulsar. See *Neutron star.*

Quasar. A blue pinpoint of light starlike in appearance (hence the name, for Quasi-Stellar Object) but with a large red shift indicating that it is far away in the expanding universe. Quasars probably are the nuclei of young galaxies going through a violent stage during or immediately following their formation. Support has been lent to this theory by the discovery of distant galaxies with bright nuclei that closely resemble quasars.

Radio. Relatively low-frequency, long-wavelength electromagnetic radiation. The universe abounds in natural radio energy, much of it produced by atoms floating in interstellar clouds and by electrons being accelerated through space by magnetic fields. See *Spectrum.*

Radio galaxy. A galaxy that emits energy unusually strong in radio wavelengths.

Radio waves. See *Radio* and *Spectrum.*

Rattail galaxy. A pair of galaxies whose interaction has released large numbers of their stars and interstellar material in the form of a pair of distended plumes or tails.

Red shift. Displacement of spectral lines in the light of stars or galaxies toward the red or low-frequency end of the spectrum. Red shifts in the spectra of galaxies have been explained as representing the velocity of galaxies moving apart in the expansion of the universe. See *Spectrum.*

Relativity, theories of. Two theories of physics created by Einstein and based in part upon the recognition that in the absence of any universally authoritative frame of reference, any observer's frame of reference must be accepted as equally valid to any other. The special theory concerns bodies in uniform motion relative to one another; it derives such consequences as the equivalence of mass and energy ($E = mc^2$) and the apparent alteration in mass, shape and the rate of passage of time of objects in motion relative to an observer. The general theory, Einstein's theory of gravitation, envisions events taking place within a four-dimensional space-time continuum; stars and planets pursue geodesics—the shortest line between two points—along the continuum. Among its many other virtues, this approach does away with the necessity of invoking a "force" of gravity.

Ring galaxy. A galaxy shaped something like a smoke ring. It appears to be a transitional period in the life of a normal spiral galaxy induced by gravitational imbalances created when it collides with a smaller galaxy.

Satellite galaxy. A small galaxy orbiting a large one. The Magellanic Clouds are the largest of the numerous satellites of our galaxy.

Seyfert Galaxy. A galaxy with an unusually bright nucleus radiating strongly in blue and ultraviolet light. About one percent of major galaxies fit this category.

SO galaxy. A galaxy shaped like a spiral galaxy but lacking spiral arms and interstellar gas and dust.

Space-time continuum. See *Relativity, theories of.*

Spectroscope. A device for breaking down light or other radiation into its component frequencies. See *Spectrum.*
10382//bed

Spectrum. Electromagnetic radiation arranged in order of its wavelength. Unlike mechanical waves, electromagnetic waves can propagate through empty space. Their wavelengths range from as much as twenty miles for long-wave radio down to 5.5 trillionths of an inch for some gamma rays. Forms of electromagnetic energy listed here in order from longer to shorter wavelengths are: radio waves, microwaves, infrared light, visible light, ultraviolet light, X-rays and gamma rays. Radio telescopes examine the electromagnetic radiation of the first two groups, optical telescopes the next three, and orbiting detectors are used for astronomical observation in X-ray and gamma ray wavelengths. In casual usage, "spectrum" most often refers to a breakdown of light; spectra are employed to analyze the composition and behavior of stars and other astronomical objects.

Spiral arm. The luminous spiral pattern in the disks of spiral galaxies that lends them their name. Spiral galaxies typically have two major arms, though these may be fragmented into exquisitely intricate patterns. See *Density wave* and *Galaxy, spiral.*

Spiral galaxy. See *Galaxy, spiral.*

Spiral nebula. A spiral galaxy. The term is an anachronism dating from the days before galaxies had been resolved into stars, when it was uncertain whether they were independent galaxies or whirlpools of gas within our own galaxy. See *Nebula.*

Star. A self-luminous body of gas sufficiently compressed for nuclear fusion to operate at its core.

Star cluster, galactic. See *Star cluster, open.*

Star cluster, globular. A spherically shaped association of stars, smaller than a galaxy. Many globular clusters are found in the halos surrounding galaxies.

Star cluster, open. An association of stars smaller, more loosely organized and younger than a globular cluster. Most open clusters are composed of stars that formed together and are destined to dissipate across space as the cluster slowly falls apart. Open clusters are found in the disks of spiral galaxies, where star formation takes place, and so are sometimes referred to as "galactic" clusters.

Supercluster. An association of clusters of galaxies. Superclusters do not appear to be gravitationally bound and so probably are being stretched out or torn apart as the expansion of the universe proceeds. See *Cluster.*

Supernova. The explosion of a star. Titanic in their force, supernovae range from ten thousand to a million times more powerful than novae. Most of the mass of the star is blasted into space, leaving behind only a dense, cinderlike core. Supernovae occur when a massive star runs out of fuel, can no longer retain the radiative pressure that has sustained it, and collapses, creating such extreme heat and pressure at the core that the star detonates like a giant thermonuclear bomb.

Supernova remnant. Material cast into space by the explosion of a star as a supernova. Often quite massive, these remnants may remain visible in optical and radio wavelengths for much longer than the few tens of thousands of years that a typical "planetary" nebula survives. See *Supernova*

Telescope. A device for gathering and focusing energy so that distant objects may be studied. Telescopes are designed in accordance with the wavelength of the radiation they are intended to collect. Large optical telescopes employ a glass mirror to bring light to focus. Radio telescopes gather the much longer waves of radio radiation with a metal dish or a mesh of wires.

Time-dilation effect. In relativity theory, the slowing of the passage of time on board a starship or other object moving close to the speed of light relative to the passage of time at the home port it left behind. Time-dilation reaches fifty percent at about ninety percent of the speed of light and increases dramatically at velocities greater than that. Enormous amounts of energy would be required to accelerate even a small ship to velocities approaching that of light.

Twenty-one centimeter radiation. Energy emitted spontaneously by free hydrogen atoms. The twenty-one centimeter wavelength lies in the radio band of the electromagnetic spectrum, at fourteen hundred twenty megacycles. Many sorts of atoms have been detected in space by virtue of the spontaneous energy emissions, but since hydrogen is the most abundant element in space, the twenty-one centimeter radiation of hydrogen is especially useful in astronomy. As the wavelength of the radiation is exact, the velocities of clouds of gas can be determined by measuring *Doppler shifts* in their radio emanations.

Ultraviolet light. Electromagnetic energy of higher frequency than visible light, lying just beyond the blue end of the visible spectrum. Extremely hot stars, such as those that have recently ejected their shells as "planetary" nebulae and are collapsing toward the white dwarf stage, are prominent sources of ultraviolet energy. See *Spectrum.*

Universe. Everything. Compare *Cosmos.*

Variable star. A star the brightness of which varies periodically. There are many sorts of variable stars, some quite useful to astronomers as distance indicators. See *Cepheid variable star.*

Violent galaxy. A galaxy producing unusually high emissions of energy. About one percent of major galaxies fit this category. Also sometimes called an exploding galaxy, a term that is misleading insofar as it suggests that the galaxy might be flying apart; at most, a violent galaxy ejects only a small portion of its mass into intergalactic space.

Visible light. See *Spectrum.*

X-rays. High-frequency, short-wavelength electromagnetic radiation. Known cosmic sources of X-ray radiation include hot clouds of intergalactic gas and putative black holes. See *Spectrum.*

Index

Italicized numbers indicate illustrations.

The text was set in Futura Book by
U.S. Lithograph, Inc., New York, New York.
The book was printed four-color offset by
Dai Nippon Printing Co., LTD., Tokyo, Japan
The book was bound by Dai Nippon Printing Co., LTD., Tokyo, Japan.

Dire Predictions
UNDERSTANDING GLOBAL WARMING

Michael E. Mann and Lee R. Kump

LONDON, NEW YORK, MELBOURNE,
MUNICH AND DELHI

Book Design Richard Czapnik
Senior Editor Steve Setford
Cartographer Ed Merritt
Production Manager Silvia La Greca Bertacchi
DTP Designer David McDonald

Original Book Design Stuart Jackman
Publisher Sophie Mitchell

Publisher Daniel Kaveney
Development Editor Erin Mulligan

First American Edition, 2009.

Published in the United States by
DK Publishing, Inc.
375 Hudson Street
New York, New York 10014

09 10 11 12 13 14 15 16 17 9 8 7 6 5 4 3

A catalog record for this book is available from the Library of Congress

ISBN 978-0-1360-4435-2

Mixed Sources
Product group from well-managed
forests and other controlled sources
www.fsc.org Cert no. SA-COC-001592
© 1996 Forest Stewardship Council

The papers used for the pages and the cover are FSC certified,
and come from N.America, where the book was printed.

The inks used throughout are vegetable inks and the special finish
on the cover is biodegradable.

OUR COMPANY is part of Pearson, a founder signatory to the UN Global
Compact. This sets out a series of principles against which we measure
ourselves in the areas of human rights, labor standards, the environment,
and anti-corruption

High-resolution workflow proofed by Media Development and Printing Ltd, UK

Printed and bound by RR Donnelley, USA

Discover more at
www.dk.com

Contents

Part 1
CLIMATE
CHANGE BASICS

Introduction

The Intergovernmental Panel on Climate Change (IPCC) was established in 1988 to evaluate the risk of human-caused climate change. Since its inception, the IPCC's periodic assessment reports have become the *de facto* conservative standard for accuracy about the scientific facts of global climate change. Unfortunately, these assessment reports, relied upon for their accuracy and often quoted by the media and scientists alike, contain high-level scientific content that can make them difficult for the general public to understand.

As the public furor over the state of Earth's climate continues to brew, it is more important than ever for informed citizens to build a basic understanding of the reasons most scientists think the global climate is in a state of crisis. However, until now, it has been difficult for interested lay readers to find reliable sources of information that are both authoritative and easy to understand.

In this book, esteemed climate scientists Michael E. Mann (who, along with other IPCC report authors, was awarded the Nobel Prize in 2007) and Lee R. Kump have partnered with the "information architects" at DK Publishing to produce *Dire Predictions*—essential reading for citizens of a world in distress. *Dire Predictions*, at just over 200 pages, presents and expands upon the findings documented in the Fourth Assessment Report of the IPCC in an illustrated, visually stunning, and undeniably powerful way for the non-scientist.

Trouble brewing
A lone lightning bolt strikes the
ground beneath an isolated
"supercell" thunderstorm at sunset.

About the IPCC

The Intergovernmental Panel on Climate Change (IPCC) was established in 1988 by the United Nations Environment Program (UNEP) and the World Meteorological Organization (WMO). The Panel was tasked with preparing a scientifically based report on all relevant aspects of climate change and its impacts, and formulating possible strategies for addressing these impacts. The self-described role of the IPCC is to "assess on a comprehensive, objective, open and transparent basis the scientific, technical and socio-economic information relevant to understanding the scientific basis of risk of human-induced climate change, its potential impacts and options for adaptation and mitigation." The IPCC strives to be policy-relevant but not policy-prescriptive.

Since its inception, the IPCC has reviewed and assessed the most recent scientific, technical, and socioeconomic information on climate change at regular intervals, periodically producing a set of comprehensive, well-documented reports. The IPCC reports summarize our continually improving knowledge of the underlying science of climate and convey the most reliable available projections for future climate change and its impacts. The reports are written by thousands of the world's leading scientists. Rigorous peer review is a hallmark of the IPCC process, and expert reviewers are called upon to comment on all aspects of the reports.

IPCC REPORTS

The information in *Dire Predictions* closely follows the findings of the IPCC Fourth Assessment Report. The authors have presented this material in a way that makes it accessible to non-scientists, and have supplemented the assessment's findings with additional and updated material.

About the authors

Dr. Michael E. Mann is a member of the Pennsylvania State University faculty, holding joint positions in the Departments of Meteorology and Geosciences, and the Earth and Environmental Systems Institute (EESI). He is also director of the Penn State Earth System Science Center (ESSC).

Dr. Mann received his undergraduate degrees in Physics and Applied Math from the University of California at Berkeley, an M.S. degree in Physics from Yale University, and a Ph.D. in Geology & Geophysics from Yale University. His research focuses on the application of statistical techniques to understanding climate variability and climate change from both empirical and climate model-based perspectives.

Dr. Mann was a Lead Author on the "Observed Climate Variability and Change" chapter of the Intergovernmental Panel on Climate Change (IPCC) Third Scientific Assessment Report published in 2001, and a reviewer for the most recent Fourth Report. He has been organizing committee chairperson for the National Academy of Sciences "Frontiers of Science" and has served as a committee member and an advisor for other National Academy of Sciences panels. Dr. Mann is the recipient of several fellowships and prizes, and has been named to the "Scientific American 50," a list of leading visionaries in science and technology. He is author of more than 100 peer-reviewed and edited publications, and is a co-founder of RealClimate.org which seeks to inform the public, journalists, and policy makers about the science of climate change.

Dr. Lee R. Kump is a Professor in the Department of Geosciences at Pennsylvania State University, and an associate of the Penn State Earth and Environmental Systems Institute, Earth System Science Center, and the Penn State Astrobiology Research Center.

Dr. Kump received his bachelor's degree in geophysical sciences from the University of Chicago and his Ph.D. in Marine Science from the University of South Florida. He is a fellow of the Geological Society of America and the Geological Society of London. In his research he uses a variety of tools, including geochemical analysis and computer modeling, to investigate climate and biospheric change during periods of extreme and abrupt environmental and biodiversity change in Earth's history.

Dr. Kump is an active researcher with over 75 peer-reviewed and edited publications. His research has been featured in documentaries produced by *National Geographic*, the British Broadcasting Corporation (BBC), *NOVA Science-Now*, and the Australian Broadcasting Corporation. He is the lead author on the preeminent textbook in Earth System Science, *The Earth System*, now in its second edition. He is on the Board of Reviewing Editors for Science as well as the editorial board of the journal *Geobiology*. He is the associate director of the Earth System Evolution Program of the Canadian Institute for Advanced Research. He also currently serves on the National Academy of Science committee for evaluating "The Importance of Deep-Time Geologic Records for Understanding Climate Change Impacts."

What's up with the weather (and the climate!)?

You have no doubt heard quite a bit over the past few years about climate change and global warming. To truly understand these terms, and to appreciate how and why human activity is causing Earth's climate to change, you need first to understand what climate is; how it differs from weather; what factors affect it; and how modern human activity is altering it. The purpose of this section of the book is to introduce you to these concepts.

Climate and weather and us

We plan our daily activities around the weather. Will it rain? Is a storm or a cold front approaching? Weather is highly variable, and, although considerable improvements in weather forecasting have been made, largely unpredictable. Climate, on the other hand, varies more slowly and is highly predictable. We know what to expect of our local climate and the climate of many familiar regions. Panama, for example, is persistently warm and wet. Residents of Siberia and northern Alaska expect long and bitterly cold winters. In the mid-latitudes, a summer day is almost certainly going to be warmer than a winter day. Climate represents the average of many years' worth of weather. This averaging process smoothes out the individual blips caused by droughts and floods, tornadoes and hurricanes, and blizzards and downpours, while emphasizing the more typical patterns of temperature highs and lows and precipitation amounts.

The reason that climate is so predictable is that it is dependent on relatively fixed features of Earth. These include Earth's spherical form, the shape of its orbit around the Sun, and its tilted axis of rotation. Other factors that determine climate have to do with the fact that Earth possesses both oceans and continents and a multi-layered atmosphere that is composed of various gases including, critically, the greenhouse gases (the importance of which we will explain in the following pages).

Climate and latitude

Radiation from the Sun plays a big role in Earth's climate. The amount of radiation Earth receives from the Sun depends fundamentally on latitude. At the equator, the Sun's rays are most directly overhead and most directly focused on Earth's surface. As we move poleward, the Sun's position at noon is lower in the sky, and so its energy is spread over a larger area, making it less intense. This is the fundamental reason

In the far north energy from the Sun is dispersed

In the tropics energy from the Sun is concentrated

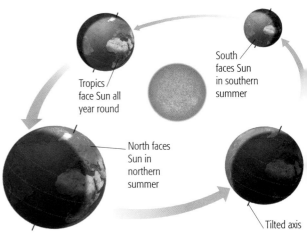

South faces Sun in southern summer

Tropics face Sun all year round

North faces Sun in northern summer

Tilted axis

that the tropics (the region between the tropic of Cancer at 23½ °N and the tropic of Capricorn at 23½ °S) are warm and the poles are so cold.

Another factor that changes with latitude is seasonal contrast: how hot the summers are, and how cold the winters are. In the tropics the difference in temperatures between summer and winter is fairly subtle, whereas at mid- to high-latitudes, the difference is quite sizable. However, the existence of the seasons themselves depends not on latitude *per se*, but on the fact that Earth's spin axis, the imaginary line that runs from pole to pole through the center of Earth, is tilted. Summer occurs in either hemisphere when the spin axis is inclined toward the Sun, while winter occurs when it is tilted away. The impact of spin axis tilt is most pronounced above the Arctic and Antarctic circles. This is why in these regions the Sun shines 24 hours a day during the summer and they remain in perpetual darkness during the winter.

Climatic bands

- Polar regions
- Temperate zones
- The tropics

Climate and the oceans

Another important factor determining continental climate is proximity to oceans. Water has a tremendous capacity for storing heat, much greater than that of the land. The oceans warm slowly during the summer and cool slowly during the winter, so coastal regions benefit from their moderating influence. In contrast, the continental interiors respond quickly to seasonal changes. This is why places like North Dakota and Saskatchewan typically have warm summers and cold winters compared to coastal locations.

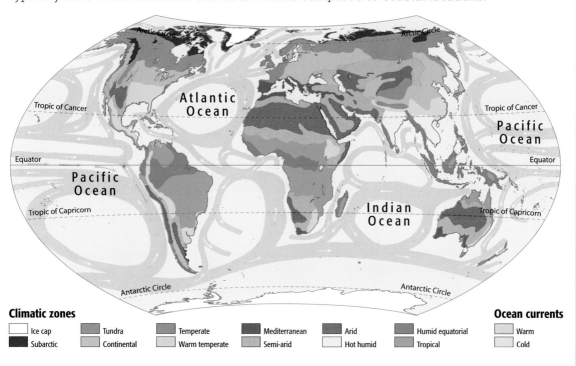

Climatic zones

Ice cap	Temperate	Arid	Humid equatorial
Subarctic	Tundra	Mediterranean	Hot humid
	Continental	Warm temperate	Semi-arid	Tropical

Ocean currents

- Warm
- Cold

Climate and the atmosphere

The existence and composition of Earth's atmosphere also influences the climate. The atmosphere is the gaseous envelope surrounding Earth; it is retained by Earth's gravitational pull. Our atmosphere features distinct layers. The first 64–80 km above the surface contains 99% of the total mass of the atmosphere and is generally uniform in gas composition with some notable exceptions, including large variations in water vapor and high concentrations of ozone, known as the ozone layer, at 19 to 50 km.

Atmospheric composition

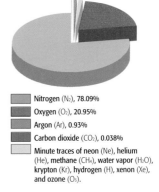

- Nitrogen (N_2), 78.09%
- Oxygen (O_2), 20.95%
- Argon (Ar), 0.93%
- Carbon dioxide (CO_2), 0.038%
- Minute traces of neon (Ne), helium (He), methane (CH_4), water vapor (H_2O), krypton (Kr), hydrogen (H), xenon (Xe), and ozone (O_3).

Exosphere
Outermost layer of the atmosphere. Extends to about 10,000 km.

Thermosphere
Extends to about 640 km.

Mesosphere
The portion of the atmosphere from about 50 to 80 km above the surface. Air becomes cooler as the altitude increases.

Stratosphere
Extends upward to a height of about 50 km. Contains atmospheric ozone layer. Temperature increases with altitude through the stratosphere, inhibiting vertical air currents, and making the stratosphere highly stable, in contrast to the troposphere.

Troposphere
Layer in contact with Earth's surface. Extends upward from the surface to about 8 km to 17 km. Air temperature decreases with altitude leading to instability. Less dense air sits below more dense air, which results in air movements and storm generation. "Weather" takes place almost exclusively within the troposphere.

Sea level

Mesopause
Boundary between the mesosphere and the thermosphere.

Stratopause
Boundary between the stratosphere and the mesosphere.

Atmospheric ozone layer
Layer within the stratosphere. Absorbs ultraviolet solar radiation so warming the surrounding atmosphere.

Tropopause
Boundary between the troposphere and the stratosphere.

OZONE LAYER

Radiation can be beneficial. Our planet would be cold without the Sun's rays, and plants need solar radiation to photosynthesize. But radiation can also be dangerous, particularly ultraviolet (UV) radiation. Fortunately, oxygen and ozone molecules in the stratosphere absorb most of the UV radiation reaching Earth. Ozone is a compound of oxygen that contains three atoms (the oxygen gas we breathe contains two oxygen atoms) and it is a lung irritant and smog producer when encountered in surface air pollution. However in the stratosphere, ozone protects life on Earth by absorbing UV radiation. In the process, the radiation destroys the chemical bonds between the oxygen atoms in the ozone molecule. Under normal circumstances, the ozone molecules rapidly reform. Unfortunately by adding chlorofluorocarbons to the atmosphere, which, like radiation, destroy ozone molecules, we've unwittingly accelerated the destruction of the ozone layer, reducing its effectiveness in protecting us from UV radiation.

Atmospheric circulation

To understand rainfall patterns, a major player in climate, we need to understand the basic principles of atmospheric circulation.

The pattern of rising moist air near the equator and sinking dry air in the subtropics is referred to as the "Hadley Circulation." The Hadley Circulation is a key component of the general circulation of the atmosphere; it helps to transport heat from the equatorial region to higher latitudes. Because of the Hadley Circulation, generally the tropics are warm and wet, while the subtropics are warm and dry. And as a result of the atmospheric circulation patterns found at higher latitudes, the mid-latitude regions experience large seasonal contrasts in temperature and rainfall patterns, while the polar regions are generally cold and dry. Rainfall in the mid-latitudes is related to the "polar front." Those of us who live in North America or Europe may know the polar front by a different name—the "storm track"—an expression that refers to the day-to-day variations in the location and intensity of the polar front.

Basic principles of atmospheric circulation

1. Water evaporates from the land and the ocean and becomes water vapor, a gas that composes part of the lower atmosphere.

2. Like a huge hot-air balloon, air near the ground in the tropics warms as a result of solar radiation, becomes buoyant, and rises.

3. As the warm tropical air rises it expands, and, like gas coming out of a spray can, it cools.

4. Cold air can hold less water vapor, so as the rising tropical air cools, water condenses out as droplets that congeal and form towering cumulus clouds and rainfall-producing thunderstorms. Thus, the tropics are rainy.

5. The rising air in the tropics draws air in from higher latitudes, forming the Intertropical Convergence Zone (ITCZ). The ITCZ migrates north and south within the tropics as the seasons change. The ITCZ is associated with trade winds that converge near the equator.

6. Air rising in the tropics moves poleward once it reaches higher altitudes. Because Earth is spinning, this poleward flow gets disrupted, and air sinks at approximately 30°S and 30°N (the subtropics).

7. This air sinking in the subtropics is now quite dry because most of the water vapor was already precipitated out of it when the air was rising. Furthermore, as air descends it gets compressed, and, like an inflating bicycle tire, warms, which dries it out even more. This is why deserts tend to occur at these subtropical latitudes.

8. A second region of rising air exists in middle-to-high latitudes (roughly 40–60°N and 40–60°S) in the region known as the polar front. Here, warm air from lower latitudes encounters cold polar air heading towards the equator. The denser polar air forces itself underneath the warmer air mass, causing it to rise, cool, and condense out its water vapor.

9. Finally, air near the poles sinks, causing the polar regions to be arid.

Cold air sinks and flows south
Northern polar front warm air rises
High-level air flows south
Ferrel cell
Low-level air flows north
Subtropical desert zone air sinks
Tropical air flows north
Hadley cell
Dry desert air flows south
Equator
Intertropical Convergence Zone (ITCZ) warm moist air rises
Hadley cell
Tropical air flows south
Subtropical desert zone air sinks
Ferrel cell
Circulation draws cool air north
Southern polar front warm air rises
Cold air sinks

GREENHOUSE GASES AND EARTH'S CLIMATE

The greenhouse effect occurs on our planet because the atmosphere (the gaseous cloud that surrounds Earth) contains greenhouse gases. Greenhouse gases are special in that they absorb heat. In doing so, they warm the atmosphere around them. When this happens, our climate changes and global warming occurs. Greenhouse gases exist naturally in Earth's atmosphere in the form of water vapor, carbon dioxide, methane, and other trace gases, but atmospheric concentrations of some greenhouse gases, such as carbon dioxide and methane, are being increased as a result of human activity. This occurs primarily as a result of the burning of fossil fuels, but also through deforestation and agricultural practices. Certain greenhouse gases, such as CFCs, and the surface ozone found in smog (which is distinct from the natural ozone found in the lower stratosphere), are produced exclusively by human activity.

Carbon dioxide (CO_2)

Water vapor (H_2O)

Ozone (O_3)

Methane (CH_4)

Nitrous oxide (N_2O)

Climate history, climate change, the greenhouse effect, and us

Looking back on Earth's history, it comes as no surprise that climates change. Indeed, on any timescale—decadal, century, millennial, or over millions of years—the climate record is anything but constant. Over the last two million years, as ice sheets advanced and retreated across northern North America and Scandinavia, climates have oscillated between very cold and more pleasant, like today. Geologists have designated this interval of time an ice age and divide it into two epochs, the Pleistocene, which lasted from 2 million years ago until 10,000 years ago, and the Holocene, which encompasses the last 10,000 years. Ice ages are marked by episodes of extensive glaciation, alternating with episodes of relative warmth. The colder periods are called glacials, the warmer periods are referred to as interglacials. The Holocene is the most recent interglacial period of Earth's most recent ice age.

Prior to the Holocene epoch, just 20,000 years ago the world was gripped by a glacial climate with ice sheets covering much of North America and Scandinavia. Prior to 2 million years ago there were no large ice sheets in the northern hemisphere, and prior to 34 million years ago there were no large ice sheets anywhere. Throughout the time of the dinosaurs (referred to as the Mesozoic Era, which spanned from 252 to 65 million years ago) the world was considerably warmer than today, and during the warmest intervals, reptiles and other cold-intolerant organisms lived above the Arctic circle. We have to go back into "deep time," more than 300 million years ago, to find evidence for a previous ice age, one that included glacial periods perhaps considerably colder and more extensive than the most recent Pleistocene glacial epoch.

Despite this backdrop of natural climate fluctuations on various timescales, human greenhouse gas emissions (see box above), are creating an atmosphere unlike anything Earth has experienced for tens of millions of years. In many respects there may be no geologic precedent for the accelerated rate of the disturbance we are imposing on Earth's climate system; the resulting impacts may be quite unlike those associated with past natural climate variation.

So how do we, as individual citizens, best address this problem? Becoming informed about the nature of the threat, and the potential solutions that are available, is a key first step. Hopefully this book, above all else, will serve to equip readers with the information necessary to make wise choices—because it is becoming increasingly clear that the decisions we make impact our collective future world.

ICE KINGDOMS

Scientists refer to the cold regions of the planet where water persists in its frozen form (i.e., regions covered with glaciers and ice sheets or with permanently frozen soils) as the cryosphere. Much of the cryosphere exists near the poles, but high-altitude mountain glaciers occur at lower latitudes. Glaciers are huge masses of ice formed from compacted snow. An ice sheet is a mass of glacier ice that covers surrounding terrain and is greater than 50,000 square kilometers. The only current ice sheets are in Antarctica and Greenland.

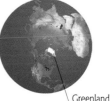

Antarctica   Greenland

The two most important regions of the cryosphere are the continental ice sheets of Antarctica and Greenland. These huge expanses of glacial ice significantly affect the amount of solar energy reflected to space, but their most significant role is their storage of water. If the ice sheets were to melt completely, sea level would rise by about 80 m. Much of this storage is in the East Antarctic ice sheet, which is less likely to be affected by anthropogenic warming in the next few centuries; West Antarctica and Greenland melting would cause a more modest but nevertheless devastating 11 m of sea-level rise. In contrast, the expansion and contraction of sea-ice (floating ice near the poles) has no effect on sea level, but can dramatically affect ocean circulation, local climate, and ecosystems. Perennially frozen ground (permafrost) influences soil water content and vegetation over vast regions and is one of the cryosphere components most sensitive to atmospheric warming trends (see p.138). Other regions of the cryosphere are also responding to climate change: the seasonal minimum sea-ice coverage of the Arctic Ocean is currently diminishing (see p.98), and most mountain glaciers are shrinking (see p.59).

Aerial view of the edge of the Greenland ice sheet
The Greenland and Antarctic ice sheets have largely survived the glacial/interval fluctuations of the last 2 million years, whereas the North American (Laurentide) and Scandinavian (Fennoscandian) ice sheets have come and gone.

Part 1
Climate
Change
Basics

 Basic principles of physics and chemistry dictate that Earth will warm as concentrations of atmospheric greenhouse gases increase. Though various natural factors can influence Earth's climate, only the increase in greenhouse gas concentrations linked to human activity, principally the burning of fossil fuels, can explain recent patterns of global warming. Other changes in Earth's climate, such as shifting precipitation patterns, worsening drought in many locations, increasingly severe heat waves, and more intense Atlantic hurricanes, are also likely repercussions of human's impact.

The relative impacts of humans and nature on climate

A variety of human actions as well as natural factors can potentially affect Earth's climate.

Natural impacts

Natural factors influencing climate include:

▪ The Sun. Over time, small but measurable changes occur in the output of the Sun—Earth's ultimate source of warming energy.

▪ Volcanic eruptions. Explosive volcanic eruptions modify the composition of the atmosphere by injecting small particles called "aerosols" into the atmospheric layer known as the stratosphere (see p.12), where they may reside for several years. These particles either reflect or absorb incoming solar radiation that would otherwise warm Earth's surface.

▪ Earth's orbit. While changes in Earth's orbit relative to the Sun influence climate on timescales of many millennia, they are not thought to play any significant role on the shorter timescales relevant to modern climate change.

Mount Pinatubo
The 1991 Mount Pinatubo eruption in the Philippines was the most explosive volcanic eruption of the 20th century. It had a cooling effect on Earth's surface for several years after the eruption.

Human impacts

The main human impact on climate is an enhanced greenhouse effect (see p.22), leading to a warming of the lower atmosphere. This is caused by increases in the atmospheric concentrations of greenhouse gases, primarily carbon dioxide produced by fossil-fuel burning (see p.26). There are also several secondary impacts of human activity. One of these involves the introduction into the atmosphere of aerosols like the ones ejected by volcanoes. These small particles (mostly sulfate and nitrate) are suspended in the atmosphere by industrial activity, such as coal combustion. Industrial aerosols reside in the lower atmosphere for only a short amount of time, and therefore must constantly be produced in order to have a sustained climate impact. The impacts of aerosols are more regionally limited and more variable than those of the well-mixed greenhouse gases. Aerosols generally reflect solar radiation back into space, and therefore represent a regional cooling influence overall. However, certain aerosols (including black carbonaceous aerosols) can, like greenhouse gases, have a surface warming influence instead. Other human impacts include stratospheric ozone depletion (see p.30) and changes in land use such as tropical deforestation, which modifies the absorptive and energy-exchange properties of Earth's surface.

Why is climate changing?

Because of the different ways that these factors influence the patterns of both solar and longwave radiation reaching Earth, it is possible to distinguish which factors are most likely responsible for any given observed change in climate. Indeed, scientists have now determined that while natural factors have been responsible for substantial changes in climate in past centuries and millennia, human impacts, increased greenhouse gas concentrations in particular, appear to be responsible for the major climate changes of recent decades (see p.34).

Industrial pollutants
The smokestacks of factories such as this paper mill spew greenhouse gases in the form of carbon dioxide and nitrous oxide. They also produce significant amounts of aerosol-forming sulfur dioxide.

Human impacts appear to be responsible for the major climate changes of recent decades.

Taking action in the face of uncertainty
The role of the scientist in global policy making

POSSIBLE PATHS OF FUTURE GLOBAL WARMING

There is considerable scientific uncertainty about how much global warming will occur and how fast it will happen, partly because the key socioeconomic factors that determine rates of fossil-fuel consumption are so unpredictable. For this reason, in the IPCC report, scientists are careful to refer to "projections" rather than "predictions" when discussing future emissions scenarios and their climatic and socioeconomic implications. Here we see the path warming has taken in the recent past and several possible projected outcomes, each of which corresponds to a different future fossil-fuel use scenario (see p.86). The right panel indicates the uncertainty associated with climate model projections for three scenarios.

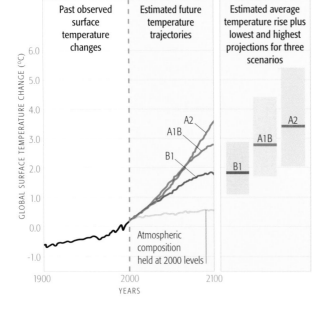

Uncertainty in global warming projections exists whether we like it or not. Some people express skepticism in response to this uncertainty, and cite it as an excuse for inaction. Scientists themselves are trained to be skeptical. They recognize that few things in science can be stated with certainty, that hypotheses can only be disproved, not proved, and that results and conclusions should be expressed in terms of this uncertainty. Unfortunately, while scientists are able to make strong conclusions from uncertain results, others view uncertainty as an indicator of flawed or inadequate scientific approaches.

Why are climate projections uncertain?

In climate science, uncertainty arises from a variety of sources, such as the inherently unpredictable nature of both the physical climate system and the human factors driving climate change, the necessary simplifications that occur when computer models are created, and incomplete knowledge about critical parameters in these models. In determining the likelihood that a conclusion is correct, climate scientists often turn to statistics, but some factors cannot be quantified by data. In these cases, likelihoods can only be established based on expert judgment. Scientific conclusions arise from time-tested theories, accurate observations, realistic models based on the fundamentals of physics and chemistry, and consensus among colleagues working in the discipline.

The Fourth Assessment Report

The Fourth Assessment Report of the IPCC presents conclusions in terms of the likelihood of particular outcomes.

These are expressed as a probability, based either on calculations or expert opinions. Likelihood ranges from virtually certain (greater than 99% probability of occurrence) to exceptionally unlikely (less than 1% probability of occurrence).

As policymakers are well aware, the risk associated with any of these projections is the combination of the probability of occurrence and the severity of the damage if it were to occur. This means that we should not ignore the projections toward the bottom of the table. For example, although it is unlikely that the Antarctic and Greenland ice sheets will collapse during the 21st century, if they were to collapse, the consequences would indeed be dire. Therefore, the risk of the ice sheets collapsing is actually quite high.

IPCC PROJECTIONS FOR THE 21ST CENTURY

This table outlines the IPCC's projections for the 21st century, ranked in decreasing order of certainty.

▪ Cold days and nights will be warmer and less frequent over most land areas ▪ Hot days and nights will be warmer and more frequent over most land areas	VIRTUALLY CERTAIN 99%
▪ If the atmospheric CO_2 level stabilizes at double the present level, global temperatures will rise by more than 1.5°C ▪ The warming over inhabited continents by 2030 will be about double the observed variability during the 20th century ▪ There will be an observed increase in methane concentration due to human activities ▪ The rate of increase in atmospheric CO_2, methane, and nitrous oxide will reach levels unprecedented in the last 10,000 years ▪ The frequency of warm spells and heat waves will increase ▪ The frequency of heavy precipitation events will increase ▪ Precipitation amounts will increase in high latitudes ▪ The ocean's conveyor-belt circulation will weaken	VERY LIKELY 90%
▪ If the atmospheric CO_2 level stabilizes at double the present level, global temperatures will rise by between 2°C and 4.5°C ▪ The future increase in global average surface temperature will be between −40% and +60% of the values predicted by climate models ▪ Areas affected by drought will increase ▪ The number of frost days will decrease, and growing seasons will lengthen ▪ Intense tropical cyclone activity will increase, with greater wind speeds and heavier precipitation ▪ Extreme high-sea-level events will increase, as will ocean wave heights of mid-latitude storms ▪ Precipitation amounts will decline in the subtropics ▪ The loss of glaciers will accelerate in the next few decades ▪ Climate change will promote ozone-hole expansion, despite an overall decline in ozone-destroying chemicals	LIKELY 66%
▪ The West Antarctic ice sheet will pass the melting point if global warming exceeds 5°C	ABOUT AS LIKELY AS NOT 35–50%
▪ Antarctic and Greenland ice sheets will collapse due to surface warming	UNLIKELY 33%
▪ The ocean's conveyer-belt circulation will shut down abruptly ▪ If the atmospheric CO_2 level stabilizes at double the present level, global temperatures will rise by less than 1.5°C	VERY UNLIKELY 10%

0 10 20 30 40 50 60 70 80 90
PROBABILITY (%)

Why is it called the greenhouse effect?

Unfortunately, the label has stuck, but the greenhouse effect in our atmosphere is not exactly like an actual greenhouse. A greenhouse lets in solar energy (mostly in the form of visible light), which keeps it warm and allows the plants inside to grow. The greenhouse stays warm primarily because its glass windows prevent the wind from carrying away the heat. This is very different from the greenhouse effect.

The greenhouse effect occurs on our planet because the atmosphere (the gaseous cloud that surrounds Earth) contains greenhouse gases. Greenhouse gases are special in that they absorb heat. In doing so, they warm the atmosphere around them. Not all gases are greenhouse gases. In fact, nitrogen and oxygen—the most abundant gases in the atmosphere—aren't greenhouse gases. Fortunately for life on Earth, which depends on some atmospheric warming to exist, other gases are, including water vapor, carbon dioxide, and methane. Without its greenhouse atmosphere, Earth's temperature would plummet to well below freezing.

We know that Earth has been a habitable planet for over 3 billion years. This means that there has always been a greenhouse effect. The carbon dioxide that humanity is adding to the atmosphere today isn't creating the greenhouse effect, it's simply intensifying it.

Hot house
The greenhouse effect does keep the planet warm like the plants inside this greenhouse, but it functions somewhat differently than a real greenhouse.

HOW THE GREENHOUSE EFFECT WORKS

Greenhouse gases allow sunlight to pass through the atmosphere and heat Earth, but they interfere with the loss of heat from the land and ocean, redirecting some of that heat back to the surface.

1 Earth absorbs solar energy and warms up.

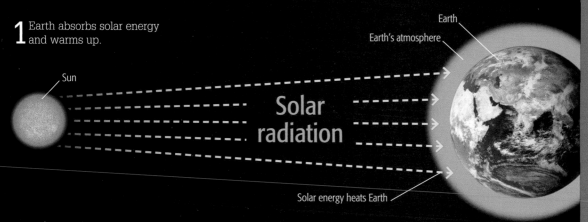

Earth

Earth's atmosphere

Sun

Solar radiation

Solar energy heats Earth

2 Like all warm objects, Earth begins to radiate heat.

Earth radiates heat, which is absorbed by its atmosphere

3 Heat radiating from Earth encounters greenhouse gas molecules in the atmosphere, and is absorbed. The atmosphere warms; as a result, it too radiates heat. Some of this heat is radiated out into space, but the rest is radiated back to Earth's surface. This extra energy warms Earth to higher temperatures. When averaged over several years, the energy radiated into space very nearly balances the solar energy absorbed by Earth. Currently, however, Earth is radiating slightly less heat into space than it is receiving from the Sun, because of the recent addition of greenhouse gases to the atmosphere. Consequently, the planet is warming.

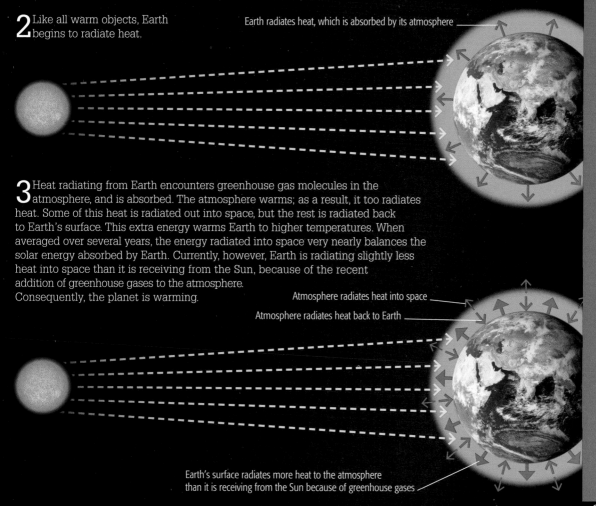

Atmosphere radiates heat into space

Atmosphere radiates heat back to Earth

Earth's surface radiates more heat to the atmosphere than it is receiving from the Sun because of greenhouse gases

Feedback loops compound the greenhouse effect

When we think about the effects of adding carbon dioxide and other greenhouse gases to the atmosphere, we have to think not just about what these gases themselves might do to climate directly, but also about their indirect effects. This is particularly important when addressing the criticism leveled by global warming skeptics that climate researchers over-emphasize the effects of carbon dioxide (CO_2), while ignoring the fact that water vapor is in fact the most powerful of the greenhouse gases. Changes in water vapor content, however, occur only primarily in response to warming or cooling, which itself is caused by changes in atmospheric CO_2. Indirect effects, such as water vapor fluctuations, are often the result of what are known as "feedback loops."

Direct radiative effect

Climate scientists refer to the "direct radiative effect" of carbon dioxide and other greenhouse gases—that is, the effect that a particular gas has on the energy budget of the planet. In terms of direct radiative effect, carbon dioxide is important, but water vapor is an even larger contributor to the overall greenhouse effect. Knowing how much indirect warming greenhouse gas emissions will cause is trickier, because of the complex feedback loops that are set in motion when greenhouse gases are added to the atmosphere.

Positive feedback

Adding carbon dioxide to the atmosphere tends to warm the atmosphere. The

POSITIVE FEEDBACK LOOP

Adding carbon dioxide to the atmosphere tends to warm the atmosphere, causing global warming.

The warm atmosphere causes surface water to evaporate and become water vapor.

Since water vapor is a greenhouse gas, the atmosphere tends to warm even more as water vapor increases.

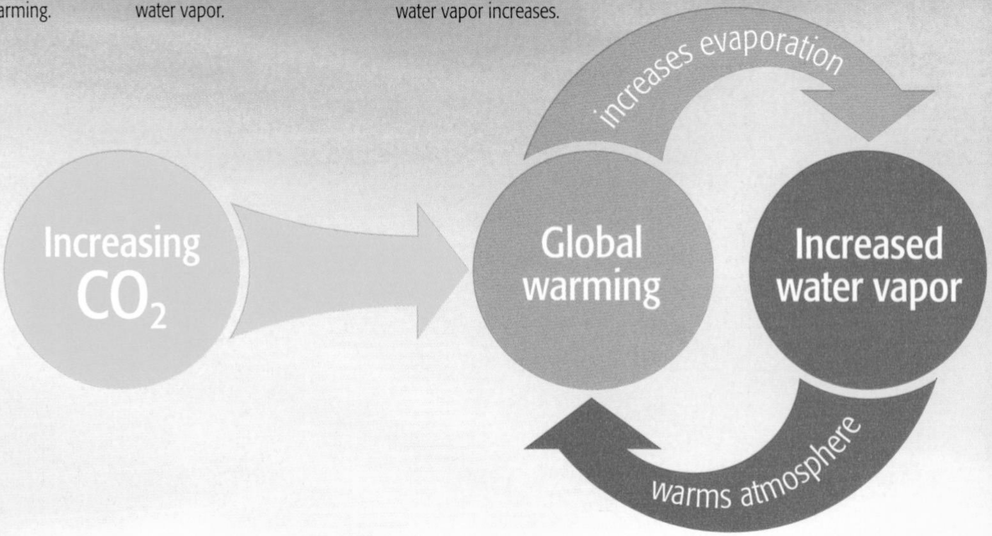

increases evaporation

Increasing CO_2 → Global warming → Increased water vapor

warms atmosphere

initial warming causes surface water to evaporate. Since water vapor is a greenhouse gas, the atmosphere will then tend to warm even more. This effect, known as the "water vapor feedback loop," is a positive feedback loop, because it amplifies the original change. Similarly, a modest amount of warming at high latitudes (e.g., Alaska and Scandinavia) can lead to a substantial melting of snow and ice, exposing the soil, rocks, or the ocean below. Because these surfaces are less reflective than snow, they absorb more solar radiation, thereby warming even more rapidly. This effect constitutes another very important positive feedback loop.

Negative feedback

Some of the additional water vapor in the atmosphere will condense to form clouds. Clouds contribute to the greenhouse effect by trapping heat in the atmosphere, but they also reflect solar energy back to space, helping to cool the planet.

Depending on where the clouds form, their overall effect therefore may be to either cool or warm the atmosphere. So things become even more complicated. If low clouds become more prevalent in response to increased CO_2, they have a cooling effect, thus offsetting some of the initial warming. This is a negative feedback effect, as it diminishes the original change.

Observations and modeling demonstrate that the overall effect of clouds is to cool the planet; this means that the warming induced by the buildup of carbon dioxide is likely to be somewhat less than it would be if the only role that water vapor played was that of an additional greenhouse gas. Nevertheless, positive feedbacks in the climate system outweigh negative feedbacks, so the expected warming from CO_2 buildup is greater than its direct radiative effect alone.

Clouds from both sides
Clouds, such as the cirrus ones shown here, can be involved in both negative and positive climate feedback loops.

NEGATIVE FEEDBACK LOOP

Adding carbon dioxide to the atmosphere tends to warm the atmosphere, causing global warming.

The warm atmosphere causes surface water to evaporate and become water vapor.

Some water vapor condenses to form clouds. Clouds contribute to the greenhouse effect by trapping heat in the atmosphere, but they also reflect solar energy back to space, helping to cool the planet.

Increasing CO_2 → Global warming → *increases evaporation* → Increased low clouds → *cools atmosphere* → Global warming

What are the important greenhouse gases, and where do they come from?

Although carbon dioxide (CO_2) has been the primary focus of concern in human-induced climate change, there are a number of other anthropogenic (human-generated) gases that also affect the radiation balance of the planet. Most of these aren't exclusively anthropogenic; with the exception of the CFCs (chlorofluorocarbons), they exist naturally. In fact, some of these gases are produced and consumed by natural processes at tremendous rates.

CARBON RECYCLING

Green plants take in carbon dioxide during photosynthesis

Rotting plants and animals return carbon to the soil

Carbon dioxide gas in air

Decomposers feed on dead plants and animals and release carbon dioxide

Burning coal and other fossil fuels release carbon dioxide into the air

The carbon cycle

Consider the CO_2 produced by your grandparents when they lit their coal stove 50 years ago. Having lain dormant in the coal for perhaps hundreds of millions of years, the carbon atoms were heated up to a high temperature in the stove, causing them to react with oxygen to produce CO_2. Let's follow a single CO_2 molecule to learn more. The CO_2 molecule in the stove escaped out the chimney into the atmosphere, where it was taken on a whirlwind tour of the planet. Sometime during its first decade of travels, the molecule entered the interior of a leaf via photosynthesis, where its two oxygen atoms were stripped away, and its carbon atom became part of the leaf. At the end of the season, the leaf fell to the forest floor. Bacteria or fungi consumed the leaf, reattaching two oxygen atoms to the carbon, and the resulting new CO_2 molecule was released back into the atmosphere. It turns out that in the 50 years since the carbon atom was freed from its lump of coal, it has been part of five different plants. This indicates that the lifetime of a CO_2 molecule in the atmosphere is about a decade. However, this cycle of uptake and release is balanced; it doesn't remove carbon dioxide, it just recycles it. Only processes acting much more slowly (over hundreds or thousands of years), including stirring into the ocean, provide a net removal mechanism.

Release without uptake

While CO_2 has always been released into the atmosphere by natural processes, fossil-fuel burning and deforestation are relatively new sources of atmospheric CO_2. Since this input hasn't been matched by a new removal mechanism, the result has been a continuous rise in atmospheric CO_2 over the last 200 years. Even if we stopped burning fossil fuels today, the return to pre-industrial levels of atmospheric CO_2 would take several centuries.

Rice paddies
Rice paddies are major methane emitters because their flooded soils provide an ideal habitat for bacteria that produce methane as a metabolic byproduct.

Bacterial byproduct

Another culprit in the human greenhouse caper is methane. Methane is a natural gas as well as an anthropogenic one. It is a metabolic byproduct of the microbes that inhabit oxygen-poor environments, such as the black mud of ponds and rice paddies, and the guts of cattle and termites. Because of the low availability of oxygen, a gas they cannot tolerate, these microbes consume organic matter but produce methane (CH_4) rather than carbon dioxide (CO_2) as a byproduct. While some organisms consume methane, most methane released into the atmosphere is removed by chemical reactions that yield CO_2. Thus an increase in methane leads to an increase in carbon dioxide. The average atmospheric lifetime for a methane molecule is about a decade. Agriculture, principally rice cultivation and livestock production, has increased the rate of methane production in recent decades, and atmospheric levels have risen correspondingly.

Agriculture is also the source of nitrous oxide (N_2O), another potent greenhouse gas. N_2O is the natural byproduct of microbes in soils and the ocean, but anthropogenic sources include nitrogen fertilizer, tropical deforestation, and the burning of fossil fuels. These human sources have increased the flow of N_2O into the atmosphere by 40–50% over pre-industrial levels; consequently, the N_2O content of the atmosphere has risen steadily. In the atmosphere, nitrous oxide is slowly broken down by sunlight; the average time an N_2O molecule spends in the atmosphere is a little over a century.

Refrigerants cause global warming

Freons, or chlorofluorocarbons (CFCs), were initially seen as a godsend, because they were efficient, non-toxic refrigerants (i.e., gases used in refrigerators). Only in the late 20th century was it realized that these gases were involved in the destruction of the ozone layer—a region of the stratosphere rich in ozone gas, which protects life on Earth from ultraviolet radiation. Unlike the other greenhouse gases, CFCs, their replacements the HFCs and HCFCs, and other similar gases used as refrigerants and fire extinguishers, have no natural source. Not only are many of these a threat to the ozone layer, they are also strong greenhouse gases.

Global warming potentials

The capacity for the various greenhouse gases to cause climate change differs because of the way in which each molecule interacts with heat. To facilitate comparison, researchers have introduced the concept of "global warming potential," or GWP. A gas's GWP is a calculation of the increase in greenhouse effect caused by the release of a kilogram of the gas, relative to that produced by an equivalent amount of CO_2. GWPs have to be expressed in terms of a time horizon, such as 20, 100, or 500 years, because the different greenhouse

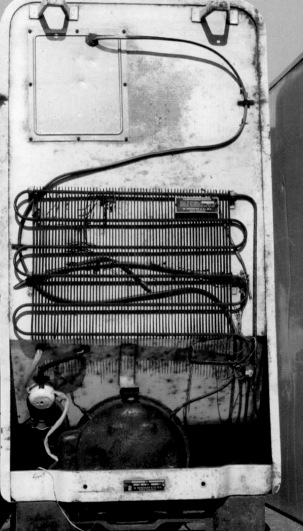

gases have quite contrasting atmospheric lifetimes. The table below illustrates the strong greenhouse capabilities of methane, nitrous oxide, and the CFCs. Fortunately, they are being emitted at much slower rates than CO_2, so their overall effect is still less than that of CO_2. Since CO_2 is involved in so many processes, it has multiple lifetimes. However, these processes recycle CO_2 rather than remove it from the atmosphere for good. Its ultimate lifetime, therefore, is considerably longer than that of the other greenhouse gases. As time passes, the relative importance of CO_2 will only increase.

AMOUNT OF GAS IN ATMOSPHERE AS EXPRESSED AS PARTS PER BILLION (ppb)

CO_2 (carbon dioxide)
Amount in atmosphere: 386,000 ppb

CH_4 (methane)
Amount in atmosphere: 1,774 ppb

N_2O (nitrous oxide)
Amount in atmosphere: 319 ppb

CFC-11 (trichlorofluoromethane)
Amount in atmosphere: 0.251 ppb

CFC-12 (dichlorodifluoromethane)
Amount in atmosphere: 0.538 ppb

HCFC-22 (trifluoromethane)
Amount in atmosphere: 0.169 ppb

LIFETIME AND GLOBAL WARMING POTENTIAL OF HUMAN-GENERATED GREENHOUSE GASES

Gas	CO_2	CH_4	N_2O	CFC-11	CFC-12	HCFC-22
Lifetime years	Multiple	12	114	45	100	12
Global warming potential						
20 years	1	72	289	6,730	11,000	5,160
100 years	1	25	298	4,750	10,900	1,810
500 years	1	8	153	1,620	5,200	549

Consider the simultaneous release of a kilogram of carbon dioxide and methane. The atmospheric lifetime of methane (CH_4) is 12 years. In the short term (the first 20 years after the gas is released), methane is a strong greenhouse gas, 72 times more powerful than CO_2. However, because it has a shorter lifetime than CO_2, on century-long timescales it becomes only 25 times as potent, and after 500 years, its potency has been significantly reduced.

Scrapped refrigerators
Although the production of CFC-11 and CFC-12 has been banned by international agreement, these gases still leak out of automobiles and from air conditioners and refrigerators decomposing in landfills.

Isn't carbon dioxide causing the hole in the ozone layer?

This is a common misconception, and confusion about the ozone layer is widespread, having permeated the media and the highest levels of government. The ozone layer is an area of high concentration of ozone molecules in the stratosphere (see p.38). The ozone layer serves an important role: it absorbs most of the solar ultraviolet radiation that bombards Earth.

Without an ozone layer, unhealthy levels of ultraviolet radiation would reach Earth's surface, making the planet largely uninhabitable.

A springtime "ozone hole"—a region where stratospheric ozone concentrations are exceptionally low—has developed in the last three decades over Antarctica. Much of the rest of the world has also experienced a reduction of the ozone layer, albeit to a lesser extent.

Although the short answer to the question of whether carbon dioxide is causing this ozone hole is "definitely not!", the more considered response points to a significant overlap between the factors causing ozone depletion and global warming.

Facts about ozone depletion and its overlap with global warming

- Global warming and ozone depletion are two global problems of paramount importance to society.

- The release and accumulation of human-generated compounds, particularly CFC-11 and CFC-12, are causing the seasonal Antarctic ozone hole and the long-term depletion of stratospheric ozone worldwide.

- CFC-11 and CFC-12 are excellent refrigerants, non-reactive chemically, and non-toxic. This non-reactivity means that they are only very slowly removed from the atmosphere by natural processes. Thus they have lifetimes of many decades in the atmosphere.

Dobson units

The reactions that destroy ozone are accelerated when particular clouds called polar stratospheric clouds (PSCs) form. These clouds form only under the coldest temperatures of the stratosphere. Although it may seem counterintuitive, the buildup of carbon dioxide in the atmospshere actually results in a cooling of the stratosphere (see p.38). This, in turn, causes more PSCs to form, thereby enhancing ozone destruction. In other words, global warming tends to promote the depletion of the ozone layer.

During their long lifetimes, CFC gas molecules can leak upward into the stratosphere, where they are fragmented by UV rays. This fragmentation releases highly reactive chlorine atoms that can destroy ozone molecules.

- CFCs are also strong greenhouse gases.

- CFCs are thus double threats to the environment. Their buildup in the atmosphere leads to the destruction of the ozone layer, and at the same time contributes to global warming.

Blue color denotes least amount of atmospheric ozone (ozone hole)

OZONE HOLE OVER ANTARCTICA
The amount of ozone in the atmosphere (measured in Dobson units) dips to very low levels in the Antarctic spring (October), because of the presence of CFCs in the upper atmosphere

Greenhouse gases on the rise

How do we know the composition of ancient air?

Although scientists have only been measuring the amount of greenhouse gas in the atmosphere for the last few decades, nature has been collecting samples for hundreds of thousands of years.

As snow accumulates on the Antarctic and Greenland ice sheets, the pressure of overlying snow compresses the snow into ice. Air trapped in the snow becomes encapsulated in tiny bubbles. Scientists drill into ice sheets, remove samples called ice cores, extract the gas from the bubbles trapped in the ice, and measure the composition of ancient air.

Scientists drill into ice sheets.

Next they remove the ice cores.

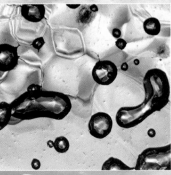

Then they extract the gas from the bubbles trapped in the ice, and measure the composition of these ancient air samples.

The impact of human activity

Together with modern observations, these analyses reveal the unambiguous human effect on atmospheric composition. As the graphs on the right demonstrate, three greenhouse gases—carbon dioxide, methane, and nitrous oxide—have been rising at dramatic rates for the last two centuries. Driven by fossil-fuel burning, deforestation, and agriculture, the recent skyrocketing trends greatly exceed the natural fluctuations of the preceding hundreds of thousands of years. Carbon dioxide and nitrous oxide have risen about 25%; methane has tripled. These gases have a powerful effect on climate, despite the fact that their concentrations are measured in parts per million (ppm) or billion (ppb). You might have to sort through millions of atmospheric molecules to find one of these molecules.

Tragically, the ice-core archive of ancient atmospheres is melting away as climates warm.

CHANGES IN GREENHOUSE GASES: ICE-CORE AND MODERN DATA

Atmospheric concentrations of carbon dioxide, methane, and nitrous oxide are shown here for the last 10,000 years. Concentrations have increased dramatically since the Industrial Revolution.

Carbon dioxide

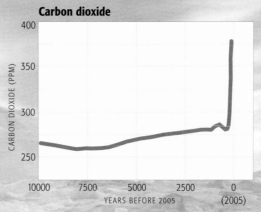

Methane

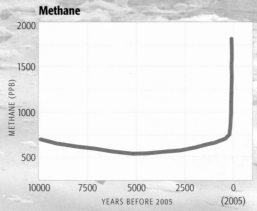

Nitrous oxide

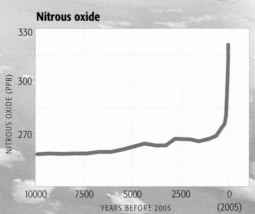

Couldn't the increase in atmospheric CO₂ be the result of natural cycles?

Some talk-radio hosts and other global warming skeptics have claimed that the undeniable rise in atmospheric carbon dioxide (CO_2) levels over the last 50 years could simply be a natural fluctuation.

How do scientists know that it is not?

There are several clues that convince scientists that the CO_2 increase is due almost entirely to fossil-fuel burning.

1 Because fossil-fuel consumption is such an integral part of the global economy, utilization rates are reasonably well known. Looking at these numbers, scientists can determine that fossil-fuel burning can more than account for the recent rise in atmospheric CO_2. In fact, the recent CO_2 rise equates to only half of what has actually been released into the atmosphere. Scientists had to probe deeper to find out what has happened to the other 50%: it has dissolved into the ocean or been taken up by the growth of forests.

No natural source for the CO₂ buildup has been identified.

2 Earth's atmosphere is naturally radioactive, because carbon-14 (radiocarbon) forms in the upper atmosphere. From measurements of tree-ring radioactivity (an indicator of atmospheric radioactivity), we know that this radioactivity remained relatively high prior to the Industrial Revolution. However, it has been decreasing over the last few decades. This indicates that much of the additional carbon driving the rise in atmospheric CO_2 levels is coming from a low-radioactivity or "radiocarbon-dead" source. Volcanoes and the deep ocean are radiocarbon-dead sources, as are fossil fuels. But we know the source couldn't be volcanoes or the deep ocean, because…

3 Carbon atoms exist in three forms, or isotopes. They all have six protons, but each has a different number of neutrons in its nucleus. The most abundant form of carbon, representing nearly 99% of all carbon, is carbon-12, an atom that has six protons and six neutrons. Carbon-12 is stable (non-radioactive); the same carbon-12 atoms that make up your body were once part of an interstellar cloud of material that congealed into our solar system nearly 5 billion years ago. A more rare form of stable carbon is carbon-13, with six protons but seven neutrons. When plants and algae photosynthesize, they preferentially use molecules of CO_2 that contain carbon-12. Thus when scientists analyze carbon sources that were derived from organic matter, like the fossil fuels coal and oil, they find that the carbon has a low ratio of carbon-13.

Just as the atmosphere has gradually become less radioactive over time, its ratio of carbon-13 to carbon-12 has been decreasing. This rules out natural, non-plant derived carbon sources, such as volcanoes and the oceans.

Conclusion

The combined trends in the atmosphere's radioactivity and its carbon-13/carbon-12 ratio are satisfactorily explained by only one source: fossil-fuel burning. Although scientists acknowledge that uncertainties exist in our knowledge of global warming, the source of the carbon that has led to the recent buildup of atmospheric carbon dioxide isn't one of them.

Radioactive trees
Tree rings record changes in atmospheric radioactivity over time.

WHAT THE NUMBERS TELL US...

The rise in atmospheric CO_2 since 1800 (graph a) is undeniable. It matches quite closely the increase in human-generated CO_2 emissions, which are quite well known (graph b). The radioactivity of the atmosphere has been decreasing (graph c), implying that the source of the increase is radiocarbon-dead. Also, the ratio of carbon-13 to carbon-12 has been decreasing, implying that the source of the increase was derived from organic matter or plants (graph d). All this points conclusively to fossil fuels as the main cause of the rise in atmospheric CO_2.

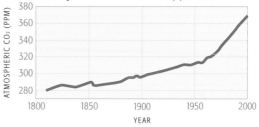

Atmospheric carbon dioxide (a)

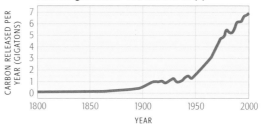

Human-generated CO_2 emissions (b)

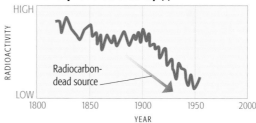

Atmospheric radioactivity (c)

Radiocarbon-dead source

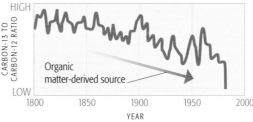

Atmospheric carbon istotope ratio (d)

Organic matter-derived source

It's getting hotter down here!
Surface temperature observations

Thermometer records have been kept for the past 150 years across much of the globe. During the last few decades, records have been kept almost worldwide. Records include surface air temperatures measured over continents and islands, and sea surface temperatures measured over oceans. By averaging these data across the globe, it is possible to estimate average global temperatures back to the mid-19th century, although uncertainties increase as we look back in time and the data become sparser. We rely on limited instrumental or historical records, supplemented by indirect evidence, to deduce temperature changes prior to the mid-19th century (see p.46).

Accelerated warming

The instrumental temperature record shows that surface warming has taken place across the oceans and land, and that the rate of warming has accelerated over the most recent decades. The average rate of global warming over the full 20th century was slightly less than 0.1°C per decade, but in the past few decades the warming rate has nearly doubled to about 0.2°C per decade. Overall, the average temperature of the globe has warmed from about 13.5°C to 14.5°C since the beginning of the 20th century.

While this warming of roughly 1°C might seem small, it is nearly one-fourth of the estimated change in the temperature of the globe between today and the depths of the last ice age, when New York City was covered by a sheet of ice almost half a kilometer thick. The warming observed so far is only a small fraction of the total

TRENDS IN GLOBAL AVERAGE SURFACE TEMPERATURE

Global temperatures have risen just under 1°C since the mid-19th century, when modern measurements began (see the red line in the graph, which indicates the average rate of change from 1860 to present times). The rate of warming has more than doubled over the past 25 years (see the yellow line).

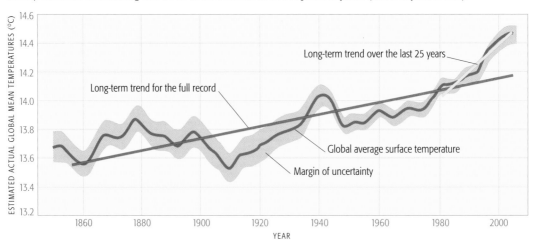

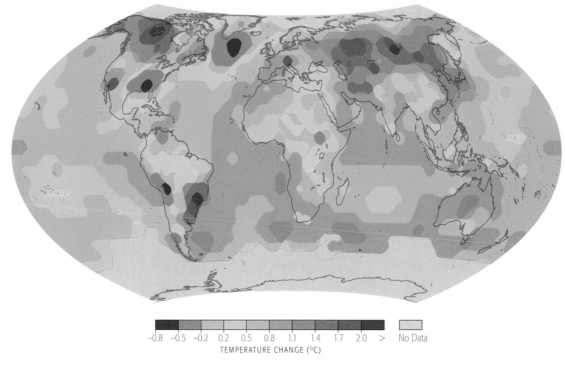

TEMPERATURE CHANGE (°C)

TRENDS IN GLOBAL SURFACE TEMPERATURE 1901–2005
Note that the pattern of warming is not uniform. Land regions, for example, have warmed more than the oceans.

warming expected during the course of the next century, if we continue to burn fossil fuels at current rates (see p.88).

No urban heat bias

Can we trust what the instrumental temperature record is telling us? It is sometimes argued that there may be an "urban heat" bias in the record, due to the fact that cities have warmed up artificially because of their high rate of energy utilization, and the dark, sunlight-absorbing properties of streets and blacktop. Since many records come from urban areas, this bias, it has been argued, may contaminate estimates of global temperature trends.

Scientists, however, have accounted and corrected for these impacts in their assessments of global temperature trends. Furthermore, similar trends are seen when only rural measurements are used.

There are other data, too, with which to counter the skeptics: thermometer measurements indicate that the ocean surface is warming significantly as well. Obviously there is no urbanization impact on sea surface temperatures.

Conclusion

Independent temperature data from the atmosphere (see p.38), the ground, and the ocean sub-surface, combined with evidence such as melting snow, ice, and permafrost (see p.98), rising sea levels, and observed changes in plant and animal behavior make it clear that Earth's surface is warming noticeably.

Is our atmosphere really warming?

At the time of the Third Assessment Report of the IPCC in 2001, there was one body of observational data that appeared to contradict the evidence for global warming. Two measurement sources of atmospheric temperatures over the past few decades—one from microwave measurements made with satellites, the other from weather balloon data—seemed to show that the lower atmosphere (the troposphere) was warming only minimally. This contradicted ground-based thermometer measurements, which indicated substantial surface warming. The inconsistency, argued the skeptics, showed either that the surface data were flawed (see p.36) and the warming trend they indicated was spurious, or that the surface warming was not caused by increased greenhouse gases, since models predicted that the troposphere will warm by as much or even more than the surface if greenhouse gases increase.

In the past few years, however, problems have been found in the older satellite and weather balloon-based assessments. The satellite estimates were compromised by errors that had artificially converted some positive trends into negative trends.

It turns out that the weather balloon data had not been sufficiently quality controlled to eliminate unreliable records.

Now that these problems have been identified and dealt with, there is considerably greater agreement between the various atmospheric temperature estimates. The corrected assessments of the satellite and weather balloon data indicate a cooling trend in recent decades in the lower stratosphere, and warming trends at the surface and in the troposphere above it. This is precisely the pattern of atmospheric temperature change predicted by climate model simulations of the response to increased greenhouse gas concentrations.

ATMOSPHERIC LAYERS

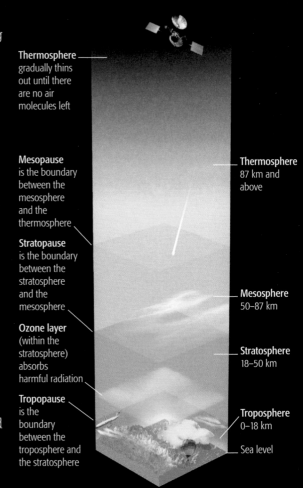

Thermosphere gradually thins out until there are no air molecules left

Mesopause is the boundary between the mesosphere and the thermosphere

Stratopause is the boundary between the stratosphere and the mesosphere

Ozone layer (within the stratosphere) absorbs harmful radiation

Tropopause is the boundary between the troposphere and the stratosphere

Thermosphere 87 km and above

Mesosphere 50–87 km

Stratosphere 18–50 km

Troposphere 0–18 km

Sea level

ATMOSPHERIC TEMPERATURE TRENDS

These graphs show observed temperature trends at various altitudes in the atmosphere. (Temperatures represent departures from the 1961–1990 average.)

ATMOSPHERIC TEMPERATURE CHANGES

This graphic shows the pattern of late 20th-century/early 21st-century atmospheric temperature changes predicted by climate models. Note that the greatest warming is observed in the tropics and in the lower atmosphere.

< −1.2 −1 −0.8 −0.6 −0.4 −0.2 0 0.2 0.4 0.6 0.8 1 1.2

TEMPERATURE CHANGE (°C)

Lower stratosphere temperature

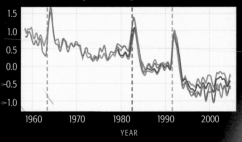

Mid- to upper troposphere temperature

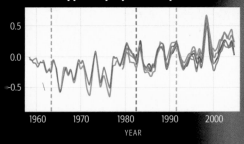

Lower troposphere temperature

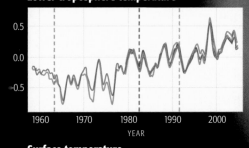

Surface temperature

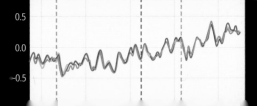

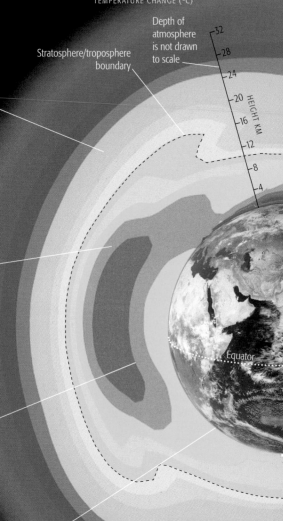

Stratosphere/troposphere boundary

Depth of atmosphere is not drawn to scale

32
28
24
20
16
12
8
4

HEIGHT KM

Equator

KEY

Air temperature analyses from thermometers, satellites, and weather balloons (°C)

Agung volcanic eruption

Back to the future
Deep time holds clues to climate change

When a doctor receives a new patient, a detailed health history is taken. Did the patient suffer these ills previously, and if so, what was the course of the illness? What are the symptoms? What brought on the illness? A physician needs to know these things before making a diagnosis.

Like your body's temperature regulatory system, Earth's climate is self-regulatory, able to resist change but subject to disturbances.

Global warming, in this context, is a planetary "fever"—not particularly high now, but possibly heading toward critical extremes.

If Earth has a planetary fever, then geologists are acting as "geo-physicians," compiling the patient's history by delving into the rock record. Rocks preserve a record of climate history, so studying rocks can tell scientists if there is any link between changing levels of atmospheric CO_2 and climate over Earth's 4.6 billion-year history.

The last two million years

Geologists are pursuing two lines of inquiry: What were the climates of the distant past, and what were the corresponding atmospheric CO_2 levels? By collecting data, geologists have been able to establish a fairly continuous record of ocean and atmospheric temperatures that spans tens of millions of years. As a result, past climate change is quite well understood. Over the last two million years, a time period referred to as the Pleistocene glacial epoch, climates have oscillated between very cold and more livable, like our current climate. By extracting air bubbles from ice cores (see p.32), scientists know that atmospheric CO_2 levels have varied in concert with these temperature swings.

Drill string
The ship's central derrick houses the drill string—thousands of meters of pipe that support the drill bit on the sea floor below.

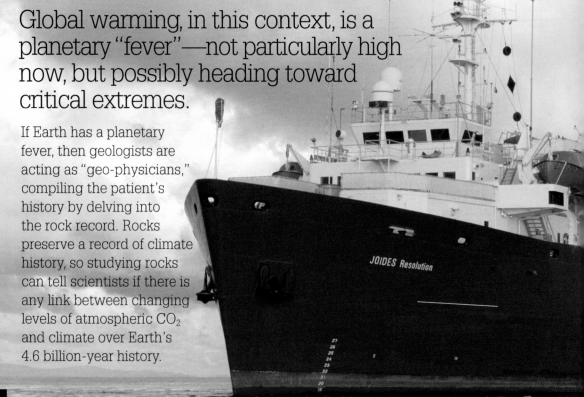

JOIDES Resolution

The last 65 million years

On longer time scales, we find that climates were generally warmer than today; the last glacial era was over 300 million years ago. The record for the last 65 million years (since the extinction of the dinosaurs) is shown below. Note that the poles were considerably warmer in the past; in the Eocene optimum, for example, alligators and sequoia forests were thriving above the Arctic Circle. The gradual cooling over the past 50 million years is curious—was it caused by declining atmospheric CO_2 levels?

To extend the record of variations in atmospheric CO_2 levels, geologists have applied knowledge from other fields, including biology, biochemistry, and soil science, to develop "proxy" measures (see p.42).

ESTIMATE OF PAST POLAR TEMPERATURE

Sediment cores show that polar temperatures 50 million years ago were up to 12°C warmer than today, and have fallen subsequently in a series of steps.

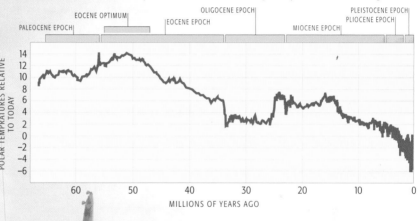

GEOLOGIC RECORDS OF CO_2 LEVELS

Estimates of atmospheric CO_2 levels, based on various proxies, show that levels have fallen over the last 50 million years.

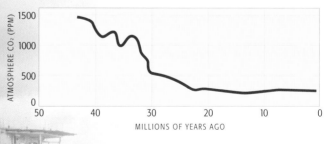

Sediment study
Scientists spend months on the *JOIDES Resolution* recovering sediment cores from the deep-ocean floor.

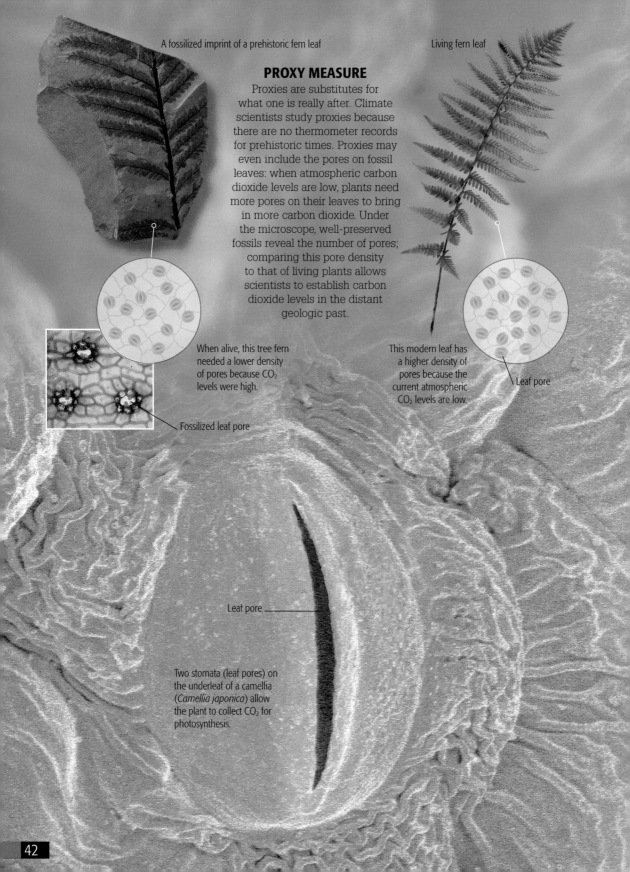

A fossilized imprint of a prehistoric fern leaf

Living fern leaf

PROXY MEASURE

Proxies are substitutes for what one is really after. Climate scientists study proxies because there are no thermometer records for prehistoric times. Proxies may even include the pores on fossil leaves: when atmospheric carbon dioxide levels are low, plants need more pores on their leaves to bring in more carbon dioxide. Under the microscope, well-preserved fossils reveal the number of pores; comparing this pore density to that of living plants allows scientists to establish carbon dioxide levels in the distant geologic past.

When alive, this tree fern needed a lower density of pores because CO_2 levels were high.

Fossilized leaf pore

This modern leaf has a higher density of pores because the current atmospheric CO_2 levels are low.

Leaf pore

Leaf pore

Two stomata (leaf pores) on the underleaf of a camellia (*Camellia japonica*) allow the plant to collect CO_2 for photosynthesis.

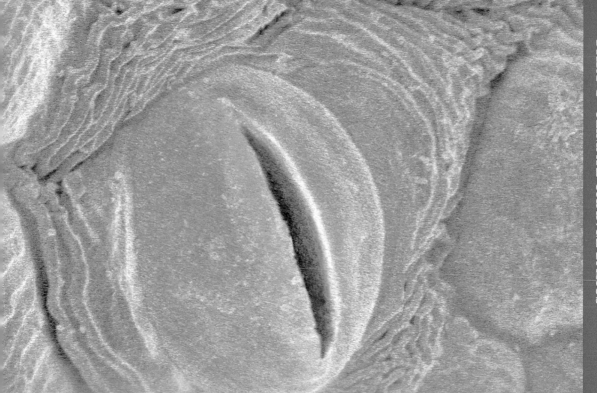

Interpreting the results

Taken together, the geologic records of climate and atmospheric CO_2 levels reveal an expected relationship: when carbon dioxide levels were high, the climate was warm, and vice versa. The correlation isn't perfect, and the mismatches are areas of current research. In particular, the cooling trend from 20 million years ago to the present doesn't seem to be reflected in a reduction in atmospheric CO_2. Are the proxies for climate and CO_2 in error during these times, or is there a third (or fourth or fifth) climate variable that we are neglecting? In this case, the growth of polar ice sheets and their high reflectivity may have provided extra cooling.

Modern measurements of atmospheric CO_2 are 386 ppm. Current estimates of available fossil-fuel reserves translate into the potential for atmospheric CO_2 to rise above 2000 ppm.

If we utilize existing fossil-fuel reserves and do nothing to capture the CO_2 released, the atmospheric CO_2 level will exceed anything experienced on Earth for over 50 million years.

But weren't scientists warning us of an imminent *ice age* only decades ago?

"Scientists ponder why world's climate is changing; a major cooling widely considered to be inevitable"

THE NEW YORK TIMES
MAY 21, 1975

A myth still perpetuated today in popular critiques of global warming science involves the supposed consensus among climate scientists in the 1970s that Earth was headed into an ice age. As this has obviously turned out not to be the case—the argument typically goes—why should we believe what scientists are saying today about global warming? As is typical with urban myths, there is a small grain of truth to this claim. Ultimately, however, the assertion is incorrect and misleading on several grounds.

It is true that decades ago climate scientists were uncertain about future trends in global average temperature, but there was no consensus that an ice age was imminent, or even that the future trend would be one of cooling. Those ideas were conveyed in sometimes quite alarmist accounts in the popular media during the mid-1970s (e.g., in *Newsweek*, *Time*, and *The New York Times*)—not in scientific publications. Indeed, the National Academy of Sciences concluded in a report published in 1975: "…*we do not have a good quantitative understanding of our climate machine and what determines its course. Without the fundamental understanding, it does not seem possible to predict climate…*"

So why all the uncertainty? First, scientists recognized that Earth was eventually due for another ice age as part of the natural cycles of cold ("glacial") and warm ("interglacial") periods that occur due to slow changes in Earth's orbit around the Sun. Just how far away the next ice age might be was not very well known. Second, scientists were already aware that human impacts on climate included both a regional cooling effect from industrial aerosols and the global warming effect of increased greenhouse gas concentrations due to fossil-fuel burning. But it was still not fully understood which of these effects would dominate in the end. We know now that the cooling from the 1950s to the 1970s was probably due to a substantial increase in the regional aerosol cooling impact, which at the time was overwhelming the greenhouse warming impact in the northern hemisphere (see p.68).

More than 30 years after the ice age scare, much has changed scientifically.

The rate of increase in greenhouse gas concentrations due to fossil-fuel burning has accelerated, while policies such as the Clean Air Acts have dramatically reduced aerosol production in most industrial regions. So the impact of increasing greenhouse gas concentrations has considerably overtaken any aerosol-related cooling. Accordingly, since the 1970s there has been even more warming than during the entire preceding century. Equally important, climate scientists now work with far more reliable models of Earth's climate system than they did 30 years ago (see p.64), and they have a better

knowledge of the various natural and human factors that influence climate. It is now clear that a natural ice age is not due for many millennia, and we know that the relative impact of aerosols has been small compared to that of greenhouse gas concentrations in recent decades. We also have considerably more data, and we know that the rate of warming in recent decades is greater than can be explained by any natural factors (see p.68).

NORTHERN HEMISPHERE CONTINENTAL TEMPERATURE TRENDS
When we compare northern hemisphere temperature trends through the current decade with the shorter record that was available in the mid-1970s (inset), we see that the trend is actually toward elevated temperatures, not cooling.

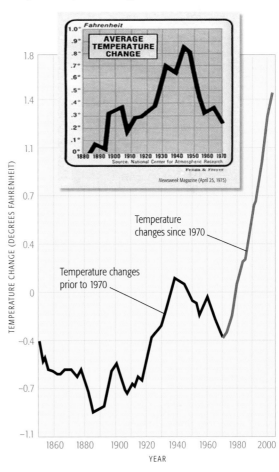

Temperature changes since 1970

Temperature changes prior to 1970

How does modern warming differ from past warming trends?

Some inaccurate accounts of Earth's climate history make reference to a period called the "Medieval Warm Period." It is sometimes asserted, for example, that because Norse explorers were able to establish settlements in southern Greenland in the late 10th century, global temperatures must have been warmer then than now. Supporters of this view also point to the fact that wine grapes were grown in parts of England in medieval times, indicating that local conditions were warmer than they are today. In fact, the ability of the Norse to maintain colonies in Greenland appears to have been related to factors other than local climate (such as the maintenance of vigorous trade with mainland Europe), and wine grapes are grown over a more extensive region of England today than they were during medieval times.

Actual scientific evidence

So how do modern temperatures compare to those in past centuries, based on the actual scientific evidence? Evidence from climate proxy data (see p.42), including tree rings, corals, ice cores, and lake sediments, as well as isolated documentary evidence, indicates that certain regions, such as Europe, experienced a period of relative warmth from the 10th to the 13th centuries, and one of relative cold from the 15th to the 19th centuries (this latter period is often referred to as the "Little Ice Age"). Other regions however, such as the tropical Pacific, appear to have been out of step with these trends.

In fact, the timing of peak warmth and peak cold in past centuries seems to have been highly variable from one region to the next. For this reason, temperature changes in past centuries, when averaged over large regions such as the entire northern hemisphere, appear to have been modest—significantly less than 1°C.

Warming everywhere

Unlike the warming of past centuries, modern warming has been globally synchronous, with temperatures increasing across nearly all regions during the most recent century. When averaged over a large region such as the northern hemisphere (for which there are widespread records), peak warmth during medieval times appears to have reached only mid-20th century levels—levels that have been exceeded by about 0.5 °C in the most recent decades.

Vineyards—in England?
Rows of grapes in the vineyards of Denbies Wine Estate, Surrey, England, UK

Simulations indicate that the peak warmth during medieval times and the peak cold during later centuries were due to natural factors, such as volcanic eruptions and changes in solar output. By contrast, the recent anomalous warming can only be explained by human influences on climate.

NORTHERN HEMISPHERE TEMPERATURE CHANGES OVER THE PAST MILLENNIUM

A number of independent estimates have been made of temperature changes for the northern hemisphere over the past millennium. While there is some variation within the different estimates, which make use of different data and techniques, they all point to the same conclusion: the most recent warming is without precedent for at least the past millennium.

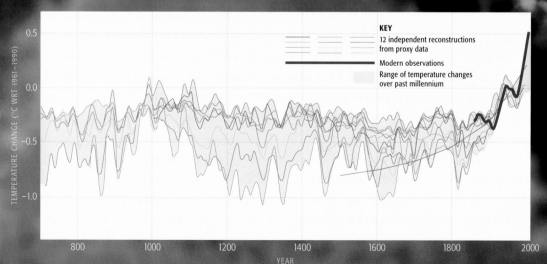

KEY

12 independent reconstructions from proxy data

Modern observations

Range of temperature changes over past millennium

What can a decade of western North American drought tell us about the future?

Much of western North America (the western US, southwestern Canada, and northwestern Mexico) has been gripped by drought since 1999. After a brief respite in late 2004, drought returned in spring 2006, and persisted through summer 2007. Lack of precipitation, reduced river flows, and lower reservoir levels all confirm the seriousness of the drought.

For much of the region, this is the most persistent and severe drought on record.

It is more widespread than the great "Dust Bowl" of the 1930s, which primarily influenced only the central US. What is particularly problematic is that the western US has entered into a pattern of severe drought just as demands for its scarce water resources are skyrocketing, due to increasing irrigation requirements by agribusiness, and dramatically growing populations in Arizona, Nevada, and Utah.

US Drought Monitor categories
- D4 Drought – Exceptional
- D3 Drought – Extreme
- D2 Drought – Severe
- D1 Drought – Moderate
- D0 Drought – Abnormally Dry

PATTERN OF US DROUGHT IN LATE AUGUST 2003
Extreme drought conditions existed across much of the western US.

WESTERN US PERCENTAGE DRY
The most recent interval of drought is unprecedented in both magnitude and duration.

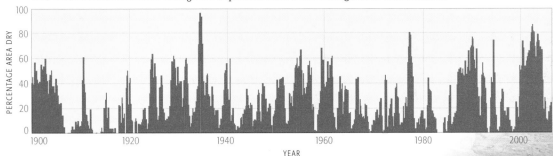

(Chart: PERCENTAGE AREA DRY (y-axis, 0 to 100) vs YEAR (x-axis, 1900 to 2000))

SEA SURFACE TEMPERATURE CHANGE (RELATIVE SCALE)

SEA SURFACE TEMPERATURE CHANGES IN THE TROPICAL PACIFIC AND INDIAN OCEANS 1998–2002

These changes are associated with the enhanced North American drought of the past decade. The concentration of warming in the western tropical Pacific and Indian oceans is reminiscent of the east–west temperature contrast typical of La Niña events.

Ocean temperature effects

Is this drought connected with human-caused climate change? That's a tricky question. Recent research ties the persistent drought conditions in western North America to a pattern of ocean surface temperatures in which the eastern and central tropical Pacific are cool relative to the western tropical Pacific and Indian oceans. This is reminiscent of the so-called "La Niña" pattern. Such a pattern has indeed persisted for most of the past decade due, in large part, to anomalously warm sea surface temperatures in the western tropical Pacific and Indian oceans. The brief cessation of drought between 2004 and 2005 was tied to an intermittent "El Niño" event, which represents the reverse pattern. In an El Niño event, the eastern and central parts of the tropical Pacific have warm sea surface temperatures, rather than the western parts. Such a pattern favors wetter conditions over much of western North America (see p.90).

Lake Mead

Lake Mead is a crucial source of fresh water for the more than 20 million people who live in the desert southwest of the US. Projections indicate that a combination of increasing water demand and worsening drought due to climate change threaten to dry up the lake in little more than a decade.

What can the European heat wave of 2003 tell us about the future?

An estimated 35,000 fatalities across Europe were attributed to a unprecedented heat wave during the summer of 2003.

Since the dramatic 2003 heat wave, climate scientists have struggled to determine to what extent this event should be attributed to human-caused climate change. With individual weather events, such as a drenching storm or a destructive hurricane, it is impossible to assess whether climate change played any role. There is simply too much randomness in day-to-day weather to make such an attribution. Heat waves, however, often take place over large areas (e.g., all of Europe or all of eastern North America), and are of extended duration

(e.g., several weeks in the case of the 2003 European heat wave). In such instances, it is possible to infer plausible connections to climate change.

In summer 2003, the maximum daily temperature records from June to August were exceeded over an extended area of Europe. In some cases, records several centuries old were broken. These record-setting conditions were part of a coherent, large-scale pattern.

2003: warmest summer in at least 500 years

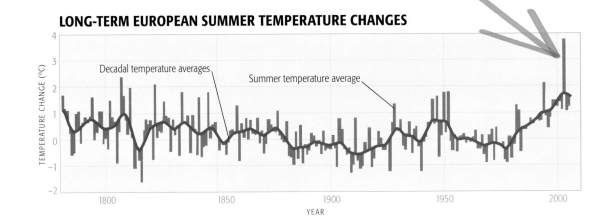

LONG-TERM EUROPEAN SUMMER TEMPERATURE CHANGES

Decadal temperature averages

Summer temperature average

TEMPERATURE CHANGE (°C)

YEAR

Temperatures for all of central Europe for the entire summer (June–August) of 2003 averaged more than 1.4°C above those for any single year on record.

Warm days

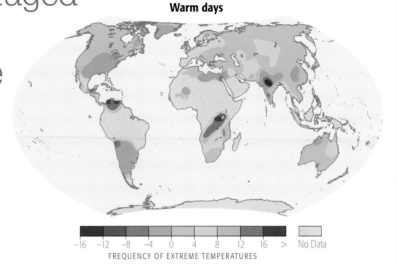

−16 −12 −8 −4 0 4 8 12 16 > No Data
FREQUENCY OF EXTREME TEMPERATURES

TRENDS IN DAILY EXTREME WARMTH

These spatial maps show the net change from 1951 to 2003 in the number of days and number of nights per decade that qualify as "extremely warm." For these purposes, "extremely warm" is defined as being in the upper 90th percentile of the full (53-year) record. Regions shaded by warm colors represent a positive trend (more extremely warm days/nights), while cold colors represent a negative trend (fewer extremely warm days/nights).

Not only was the summer of 2003 the warmest since reliable instrumental records began in the late 1700s, but more uncertain longer-term historical records suggest that the summer was the warmest in at least 500 years. This event did not occur in isolation. Instead, it is part of a long-term trend towards more frequent daily extremes of daytime and nighttime warmth over most of the world's major landmasses.

Warm nights

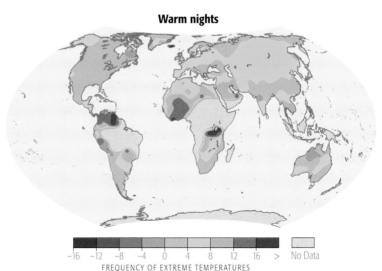

−16 −12 −8 −4 0 4 8 12 16 > No Data
FREQUENCY OF EXTREME TEMPERATURES

The 2003 European heat wave was associated with an extreme poleward expansion of the high-pressure subtropical zone. This situation is predicted to become more common with human-caused climate change (see p.100).

SUBTROPICAL ZONE EXPANSION

There are two subtropical zones. They lie just to the south of the northern polar jet stream, and to the north of the southern polar jet stream. Climate change is predicted to lead to a poleward migration of the polar jet streams, allowing the dry subtropical zones to penetrate further into mid-latitude regions such as Europe and the US during the summer season.

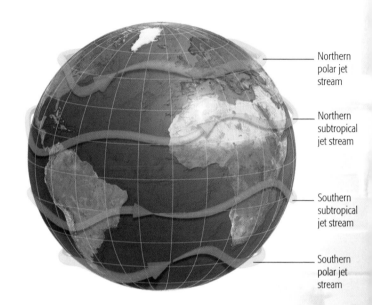

Northern polar jet stream

Northern subtropical jet stream

Southern subtropical jet stream

Southern polar jet stream

Model simulations indicate that the 2003 event would have been highly unlikely to occur in the absence of human-induced climate change.

In fact, such a likelihood amounts to a roughly once-in-a-thousand-years occurrence. By contrast, when human-induced warming is taken into account, the event is predicted to occur at least once in any given century on average. While this doesn't prove that global warming caused the 2003 European heat wave, it underscores how the increasing occurrence of heat waves is consistent with theoretical expectations, and gives us a taste of what is likely to be in store with future climate change.

Hot in the city
Tourists cool off in London's Trafalgar Square during the 2003 European heat wave.

A tempest in a greenhouse
Have hurricanes become more frequent or intense?

One of the most contentious issues in climate research involves the impacts of climate change on tropical cyclone and hurricane behavior.

A hurricane is defined as an Atlantic tropical cyclone in which winds attain speeds greater than 119 kmh.

Since modern observations of hurricanes from satellites and aircraft reconnaissance missions are only available for the past few decades, it is difficult to determine whether long-term hurricane trends exist, and whether those trends are associated with human-caused climate change.

One measure of tropical cyclone activity is "accumulated cyclone energy," which takes into account the total energy of storms over their lifetime. Such measurements for the last few decades indicate increases in tropical cyclone activity over most, but not all, major formation basins. In some cases, alternative measures of tropical cyclone activity that quantify the destructive potential or "powerfulness" of storms show dramatic recent increases. These increases appear to be closely related to rising sea surface temperatures in the tropical Atlantic (the only region for which long-term records are available).

There is also evidence of a trend toward greater numbers of tropical cyclones in the tropical Atlantic over the past two decades. A global trend is less clear,

The eye of the storm
Hurricane Katrina formed on August 23, 2005. It was the costliest hurricane in U.S. history, wreaking havoc along the Gulf Coast from Lousiana to Alabama, and devastating the city of New Orleans. Katrina was also among the deadliest hurricanes in U.S. history, and was one of three Category 5 Atlantic hurricanes that formed during the record-breaking 2005 season.

however, when the numbers are averaged over all of the major hurricane basins.

Basic theoretical considerations as well as detailed climate-model simulations indicate a likely increase in the average intensity of tropical cyclones and hurricanes in all major formation basins (see p.103). That said, it is uncertain how the actual number of tropical cyclones is likely to be affected by climate change. Accordingly, this remains an area of active climate research.

SEA SURFACE TEMPERATURE VS POWERFULNESS IN TROPICAL ATLANTIC CYCLONES

Sea surface temperatures are closely related to the powerfulness of tropical cyclones. The dramatic increase in cyclone powerfulness over the past decade closely parallels the anomalous rise in sea surface temperature.

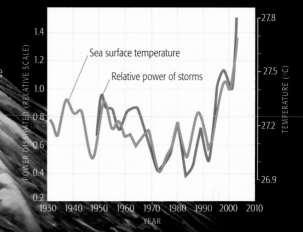

ACCUMULATED CYCLONE ENERGY IN TROPICAL CYCLONE-PRODUCING BASINS

There is a clear upward trend in cyclone energy for the tropical Atlantic in recent years. Trends are less clear for cyclones in the Pacific and Indian oceans. Straight red lines represent the long-term average values of accumulated cyclone energy (ACE) over the 1981–2000 period.

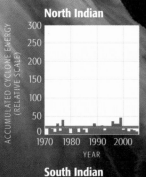

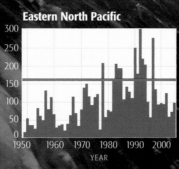

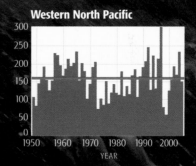

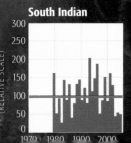

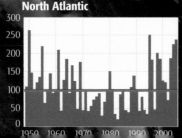

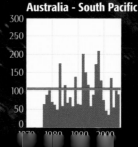

The vanishing snows of Kilimanjaro
An icon of climate change?

There is perhaps no snow-capped peak in the world as iconic as Mount Kilimanjaro, Tanzania, which was immortalized in Ernest Hemingway's *The Snows of Kilimanjaro*. A remarkable group of perennial ice fields can be found at altitudes of roughly 5 km and above atop Mount Kilimanjaro, at a latitude only a few degrees south of the equator.

Like mountain glaciers in many parts of the world, the snows of Kilimanjaro provide a key source of fresh water. The ultimate demise of Kilimanjaro's ice, projected to take place sometime within the next two decades at current rates of decline, consequently poses a threat to the people who inhabit the region.

So are the snows of Kilimanjaro a victim of global warming?
We know that mountain glaciers the world over are disappearing, and that this disappearance is generally related to increased melting due to warmer atmospheric temperatures. In fact, at the altitudes where most tropical mountain glaciers are found, the atmosphere has warmed even more in recent decades than Earth's surface.

1912 (Average area of snow: 12 km²)

1970 (Average area of snow: 5 km²)

2000 (Average area of snow: 2.5 km²)

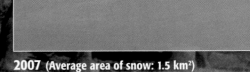

2007 (Average area of snow: 1.5 km²)

Kilimanjaro has had permanent ice fields for 12,000 years: it is unlikely that their current dramatic retreat is a coincidence.

However, it is not as simple as saying warmer temperatures cause more ice to melt. While some melting has been observed recently on Kilimanjaro, ice loss at these latitudes and elevations occurs mostly through "sublimation" (the evaporation of ice directly into the atmosphere), rather than by melting. The rate of sublimation is influenced by factors in addition to temperature, such as cloud cover and humidity. Also, the amount of ice that exists on mountain glaciers isn't controlled by melting or sublimation alone, but instead represents a balance between the rate of ice loss due to those processes, and the rate of accumulation of ice. Accumulation is determined by the amount of precipitation that falls each year, so it is likely that decreased snowfall in the region has had a significant impact on the extent of the ice fields. The changing precipitation patterns, leading to drier conditions in the region, are tied to larger-scale climate changes. In this sense, human influence on climate is probably responsible for the imminent demise of Kilimanjaro's snows, even if warmer regional temperatures alone are not.

Mount Kilimanjaro
Kilimanjaro, in Tanzania, Africa, is one of a number of locations around the world where ancient mountain glaciers are disappearing before our eyes.

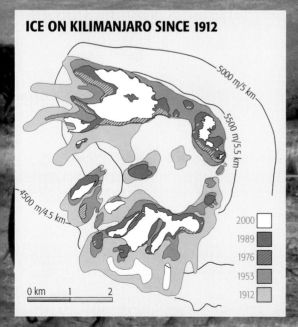

ICE ON KILIMANJARO SINCE 1912

5000 m/5 km

5500 m/5.5 km

4500 m/4.5 km

2000	
1989	
1976	
1953	
1912	

0 km   1   2

The day after tomorrow
Will the ocean's conveyor belt shut down?

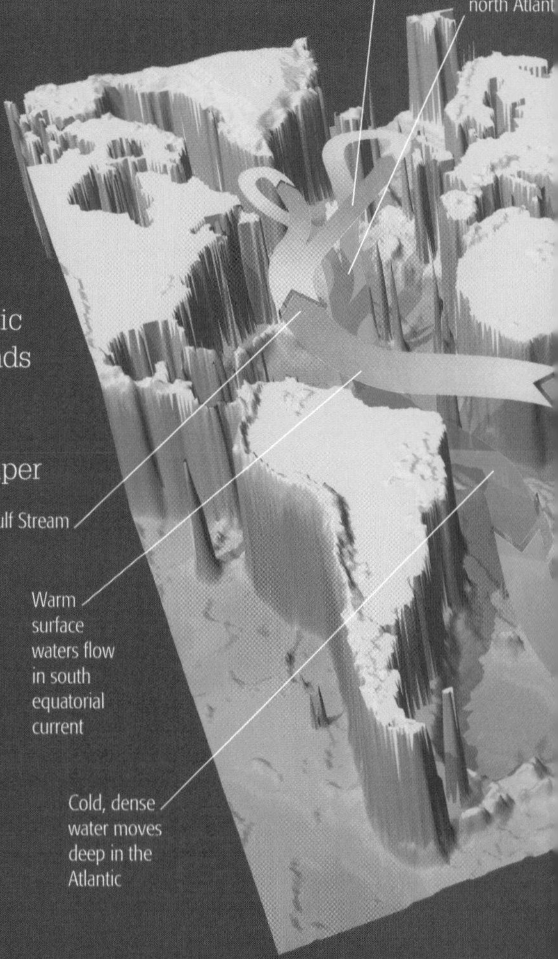

In the 2004 blockbuster movie *The Day After Tomorrow*, global warming causes catastrophic melting of the polar ice sheets. This, in turn, leads to changes in ocean salinity and a shutdown of ocean circulation. The cooling that ensues when ocean waters quit circulating causes super cyclones that wreak havoc worldwide.

Was the scenario in *The Day After Tomorrow* wholly the brainchild of a screenwriter, or does it have scientific underpinnings?

North Atlantic Drift

Cold, salty w downwells north Atlant

Gulf Stream

Warm surface waters flow in south equatorial current

Cold, dense water moves deep in the Atlantic

Although the rate and scale of melting and its consequences are exaggerated in the movie, and in some cases preposterous, the basic notion that the ocean's conveyor-belt circulation is sensitive to the salinity of the North Atlantic is rooted in scientific theory and observation.

If global warming were to increase rainfall in the North Atlantic or cause significant melting of the Greenland ice sheet, the North Atlantic would become less salty. Less-salty water might not sink, which could hamper the flow of the warm ocean current known as the Gulf Stream off the east coast of the United States– and as the North Atlantic Drift along its northeasterly extension (see figure)–and affect global water circulation patterns.

It may have happened before
The climate of northern Europe is milder than that of northern North America at an equivalent latitude. This is the case because air masses move across the Atlantic toward Europe and are warmed by the relatively warm ocean below. A shutdown of the conveyor belt would limit the flow of the warm North Atlantic Drift, cooling northern Europe.

Diffuse warm and cold water upwells in north Pacific Ocean

Warm north equatorial current flows at the surface in the central Pacific

CONVEYOR-BELT CIRCULATION

The Gulf Stream and North Atlantic Drift (represented by the red arrow in the North Atlantic part of this image) carry warm, salty water into the northernmost Atlantic. As this water moves northward it cools, releasing heat to the atmosphere. It also becomes saltier as water evaporates from the surface. Cooler and saltier water is denser, so eventually it sinks (the blue arrow represents the water that has cooled and sunk). At the surface, more warm Gulf Stream water is drawn northward to replace it. The pattern of the ocean's conveyor-belt circulation involves water rising to the surface in the Pacific and Indian oceans, where it warms and once again becomes the Gulf Stream; the actual situation is a bit more complicated, but beyond the scope of this book. One circuit of the global conveyor belt takes 500–1000 years.

Diffuse upwelling occurs in the Indian Ocean

Warm equatorial surface current flows through Indonesian archipelago

Combined mass of cold water moves slowly deep around Antarctica

Cold, dense water flows north deep into the Pacific Ocean

Atlantic water is joined here by more cold water from near Antarctica

Paleoclimatologists have documented an abrupt climate change that occured the end of the last ice age, 12,000 years ago. They think it can be attributed to a temporary shutdown of the North Atlantic ocean circulation. This event was characterized by an abrupt return to glacial climate conditions in northern Europe. The likely cause is a massive infusion of glacial meltwater, which caused a shutdown of the ocean's conveyor-belt circulation.

Climate models predict that under most climate change scenarios, the Greenland ice sheet may be devastated in the coming centuries. The rate of its melting is most uncertain, and it is dependent on the nature of poorly known feedback loops within the ice-sheet system that could accelerate the melting dramatically (see p.98). Even so, the latest climate models indicate that the rate of injection of fresh water into the North Atlantic will not be sufficient to cause a complete shutdown of the conveyor-belt circulation. Some slowdown will likely occur, though, and the resulting cooling of the North Atlantic and northern Europe might offset some of the effects of global warming. The real "Day After Tomorrow," albeit bleak, will bear little resemblance to the Hollywood version.

The last interglacial
A glimpse of the future?

Driving south on highway U.S. 1 from Miami, Florida, you pass a road sign for the Windley Key Fossil Reef Geological State Park, the site of a former limestone quarry. Here, you are at the highest point in the Florida Keys—islands that 125,000 years ago were an impressive chain of coral reefs. At that time, sea level was 6 m above where it is today, most likely because the Greenland ice sheet was smaller and much of the water it comprises today was at that time in the ocean, not locked up in glacial ice. The climate then began to cool, and the reefs became exposed as the sea level fell when water from the ocean evaporated and froze to form ice sheets in the northern hemisphere. At the height of the last glaciation, the seas had retreated 10 km, and the Keys stood 120 m above the ancient sea level. Now they are barely a few meters above sea level, with Windley Key at the highest elevation (6 m). These fluctuations in sea level are the result of the 40-thousand- to 100-thousand-year "glacial–interglacial" cycles of the last 2 million years. These cycles occur in conjunction with the repeated growth and demise of the almost 2-km-thick North American ice sheets that covered most of Canada and the northern United States, as well as other smaller ice sheets in northern Europe.

Will the Florida Keys turn back to coral reefs once again?

Climate in the last interglacial
Indeed, with modern global warming causing sea levels to rise again, low-lying areas like the Florida Keys are gradually disappearing (see p.98). To understand the future implications of these changes, scientists naturally turn to time periods with analogous conditions—like the last interglacial—for clues to questions such as how much the Greenland ice sheet is likely to shrink.

Florida Keys
Highway U.S. 1 extends over a fossil coral reef that formed 125,000 years ago.

A wealth of information on the climate of the last interglacial has been collected in recent decades. Ice cores (see p.82) reveal that atmospheric CO_2, methane, and nitrous oxide were then close to pre-industrial levels. Coastal ocean temperatures were generally warmer than today, there was considerably less sea ice surrounding places like Alaska, and boreal forests had overtaken regions that are now tundra in Siberia and Alaska. Arctic summers were warmer than pre-industrial summers by 5°C.

Earth's orbit affects temperature

Why was the last interglacial so warm— warm enough to melt the Greenland ice sheet? We know from ice-core data that it wasn't due to higher CO_2 levels. The answer is that the northern hemisphere was receiving 10% more solar radiation than it does today, not because the Sun was brighter, but because Earth's orbit around the Sun was different than it is today. Earth's orbit changes slowly and regularly in response to the tug of the Sun, Moon, and the large planets, especially Jupiter. The orbital configuration 125,000 years ago provided more direct rays to northern latitudes in the summer.

Even though this wasn't a greenhouse warming event, and thus not a direct analogy, the information we can learn from the last interglacial suggests that the Greenland ice sheet is subject to considerable shrinkage from relatively subtle changes.

EARTH'S CHANGING ORBIT

Earth's orbit and rotation are not constant, but change cyclically over many millennia. Its orbit varies from elliptical to circular, and the planet also tilts about its axis and wobbles as it rotates. Over time, these changes affect temperature, and they can be correlated to climate swings and glacial–interglacial cycles.

Orbital ellipticity (100,000-year cycles)

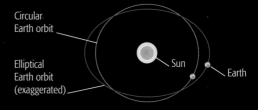

Circular Earth orbit

Elliptical Earth orbit (exaggerated)

Sun

Earth

Tilt (41,000-year cycles)

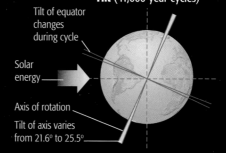

Tilt of equator changes during cycle

Solar energy

Axis of rotation

Tilt of axis varies from 21.6° to 25.5°

Wobble (19,000- and 23,000-year cycles)

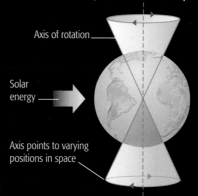

Axis of rotation

Solar energy

Axis points to varying positions in space

How to build a climate model

Building a model of Earth's climate is a very challenging endeavor; the climate is governed by many complex physical, chemical, and biological processes and their interactions. Earth's climate can be thought of as a system in which fluids with very different properties (Earth's oceans, atmosphere, and ice "cryosphere") interact with each other, as governed by the laws of fluid dynamics, thermodynamics, and radiation balance. There are many further complications, however, that must be taken into account in modeling Earth's climate. For example, life on Earth (the "biosphere"), which includes both plants and animals, plays a key role. The biosphere is involved in the global recirculation of water and carbon, and it influences the composition of the atmosphere and Earth's surface properties—all of which impact on climate.

Simple climate models

The simplest models ignore the three-dimensional structure of Earth, atmosphere, and oceans, and simply focus on the balance between incoming solar energy and outgoing terrestrial ("heat") energy. It is the balance between these incoming and outgoing sources of energy that determines temperatures on Earth. Even in these simple models, the greenhouse effect (see p.22) must be accounted for. This is usually accomplished through a modification that represents the way heat is absorbed and emitted by the atmosphere. It is also essential, even in simple models, to account for "feedback loops" that can either amplify (positive feedback) or diminish (negative feedback) the impacts of any changes (see p.24). In most climate models, the net impact of feedbacks roughly doubles the magnitude of the expected warming or cooling response to imposed changes.

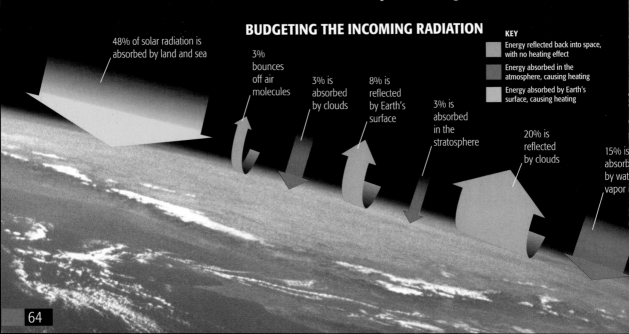

BUDGETING THE INCOMING RADIATION

48% of solar radiation is absorbed by land and sea

3% bounces off air molecules

3% is absorbed by clouds

8% is reflected by Earth's surface

3% is absorbed in the stratosphere

20% is reflected by clouds

15% is absorb by wat vapor

KEY

Energy reflected back into space, with no heating effect

Energy absorbed in the atmosphere, causing heating

Energy absorbed by Earth's surface, causing heating

The most complex climate models, referred to as "General Circulation Models," take into account the full three-dimensional structure of the atmosphere and oceans, the arrangement of the continents, the details of coastlines and ocean basins, and the surface topography. These models calculate not only surface temperatures, but also other important climate variables, such as precipitation, atmospheric pressure, surface and upper level winds, ocean currents, temperatures, and salinity. All this is accomplished by breaking the oceans and atmosphere into many small grid boxes, and using the underlying physical, chemical, and biological relationships to calculate values for the properties of each box and the interactions between the different boxes.

Should climate model predictions be trusted?

Current climate models do a remarkably good job of reproducing key features of the actual climate such as the jet streams in the atmosphere, the seasonal band of rainfall and cloudiness that migrates north and south of the equator, and even the complex internal climate oscillation associated with the El Niño phenomenon (see p.90). They also closely reproduce past changes (see p.68). We therefore have good reason to take their predictions of possible future changes in climate seriously.

Atmosphere is divided into 3-D grid boxes, each with its own local climate

Air in grid boxes interacts horizontally and vertically with other boxes

Influence of vegetation and terrain is included

Water in oceanic grid boxes interacts horizontally and vertically with other boxes

Oceanic grid boxes model currents, temperature, and salinity

COMPLEX CLIMATE MODELLING

A climate model is a computerized representation of Earth including its atmosphere, oceans, and various other components, based on a three-dimensional global grid. The model then applies these changes to its virtual world, to see what effect they have on the climate.

LOST ENERGY

Only 48% of incoming solar energy reaches Earth's surface to heat the continents and oceans. Nearly one-third of the total energy that encounters the atmosphere is immediately returned to space—reflected by clouds or air molecules. Additional radiation is reflected by Earth's surface, especially in icy regions. The remainder is absorbed by stratospheric gases and tropospheric clouds and water vapor.

Profile:
James Hansen

James Hansen is a well-known climatologist and Director of NASA's Goddard Institute for Space Studies (GISS), a major climate-modeling center. Hansen, a member of the prestigious U.S. National Academy of Sciences, came to prominence during his congressional testimony in the hot summer of 1988.

On June 23 of that year, in front of the US Congress, Hansen became the first scientist to publicly testify that society was already beginning to witness the effects of human-caused climate change. His testimony appears prescient when we look at what has happened since his now-famous pronouncement.

Projections proved correct

During that testimony, Hansen presented projections about likely future warming in terms of three possible scenarios (see chart at right).

The projections Hansen made in 1988 have proven to be a key validation of the models used by climate scientists.

Comparing Hansen's predictions of future global temperature changes with actual temperature observations reveals a remarkable match. This shows that successful projections of global temperature can be made decades in advance.

In 2001, Hansen received a Heinz Award in the Environment for his "exemplary leadership in the critical and often-contentious debate over the threat of global climate change." He remains a productive climate scientist, and is actively involved in efforts to educate the public and policy makers about climate change and the threat it poses to society.

HANSEN'S THREE PROJECTED GLOBAL WARMING SCENARIOS

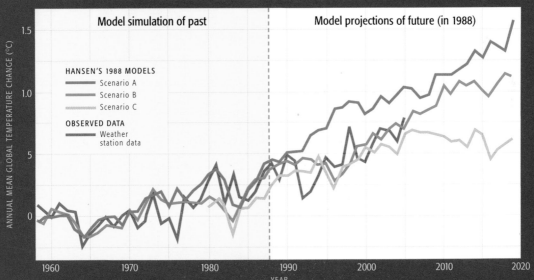

This graph shows climate model simulations performed by James Hansen in 1988, updated to indicate how projected global average surface temperatures have compared with the actual observed temperatures. Scenario B (the blue line) corresponds roughly to the actual greenhouse gas emissions pattern that society has followed since 1988.

Comparing climate model predictions with observations

The average annual temperature of the planet is not expected to be constant in the absence of human influence. Variations in solar energy input warm and cool the planet on yearly, decadal, and longer timescales, while volcanic eruptions cool the planet from months up to years after a major eruption.

Using climate models

We can use climate models to calculate how natural variations in solar energy and volcanic aerosols alone would have driven climate change over the last

century if there hadn't been any human influence (graph 1), and compare the result to the observed record (graph 2). Then we can include the human influence (graph 3), and see if the model predictions fit the observed data better (graph 4). The temperature deviations in the graphs below are expressed relative to the average temperatures from 1901 to 1997. The good fit between actual observations and models that take into account human actions and actual data gives us confidence that we can predict future climate responses to fossil-fuel burning.

PREDICTED/OBSERVED CLIMATE TRENDS

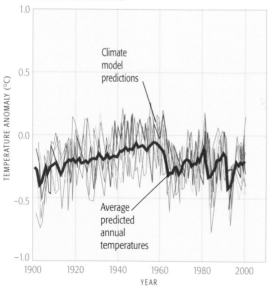

1 **Predicted temperature trends from models, taking into account the impacts of natural forces alone**

Thirteen different climate models indicate which portion of the annual average temperature variations over the last century can be attributed to natural forces alone.

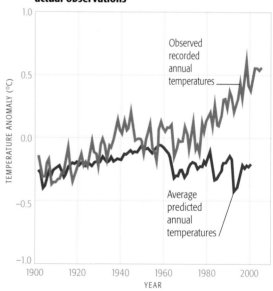

2 **Comparison of the average of the model results in graph 1 to actual observations**

The warming trend of the last 40 years is not reflected in the models that take into account the impacts of natural forces alone.

**Mount Pinatubo
June 1991**
Pinatubo began errupting in the Philippines on June 12, 1991. It caused a subsequent 0.5°C cooling of the atmosphere.

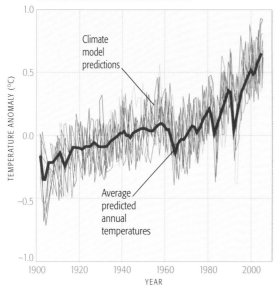

3 Predicted temperature trends from models taking into account the impacts of both natural and human forces

Climate model predictions

Average predicted annual temperatures

TEMPERATURE ANOMALY (°C)

YEAR

When human forces, especially the buildup of atmospheric greenhouse gases and industrial aerosols, are applied to these same climate models, the models shift to predict a gradual warming over the last several decades.

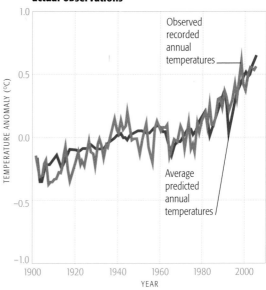

4 Comparison of the average of the model results in graph 3 to actual observations

Observed recorded annual temperatures

Average predicted annual temperatures

TEMPERATURE ANOMALY (°C)

YEAR

These model results clearly show that both natural and human forces impact climate.

Regional vs global trends

An important development in the latest IPCC report is that it is now possible to tie temperature changes over the oceans, bodies of land, and at the regional scale of individual continents to human activity. When scientists compare observed and modeled temperature changes at these regional scales, they find that the warming trends observed for individual continents, such as North America, Europe, and Africa, cannot be explained by natural factors like volcanoes and changes in solar output. As we have seen on a global scale (see p.68) only when the models include the human, or "anthropogenic," component—warming due to increased greenhouse gas concentrations and the more minor cooling impact of industrial aerosols—can they explain observed regional warming trends (see graphs at right). This indicates that human influences are now having a detectable impact on temperature changes measured in individual regions.

Temperature changes, however, are just one of the many regional impacts of climate change. Influences on other climate phenomena, such as drought and rainfall, may represent even more significant human impacts (see p.89).

Blue planet
This view of the Earth was taken from the space shuttle *Endeavour*.

GLOBAL TRENDS: LAND VS OCEAN

These graphs compare actual observations and model results both with and without human factors included. Only the models that take into account both human and natural factors make predictions that look like actual data trends.

Global

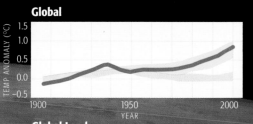

Global Land

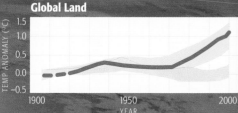

Global Ocean

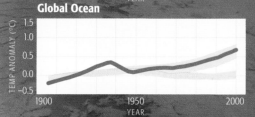

KEY

Predicted by models taking into account natural factors only

Predicted by models taking into account human and natural factors

Actual observations

Spacial coverage less than 50% (i.e., data is available for only 50% of the area in question for the indicated time interval)

REGIONAL CONTINENTAL TRENDS

These graphs compare actual observations and model results both with and without human factors included. As on the global scale, only models taking into account both human and natural factors make predictions that correspond to actual data.

North America

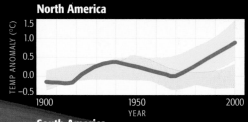

South America

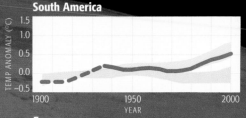

Europe

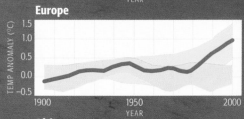

Africa

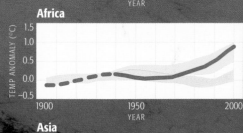

Asia

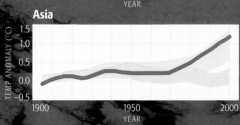

Australia

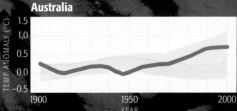

"Fingerprints" distinguish human and natural impacts on climate

Not all factors that affect climate have the same pattern of influence on temperature. One way to distinguish natural and human sources of modern climate change is to look for particular "fingerprints"—that is, specific spatial patterns of change that help to identify the likely underlying factor associated with that change.

Natural impacts

Two natural factors have influenced climate change over the past century:
- Changes in solar output
- Explosive volcanic activity

Human impacts

Major human impacts on climate include the greenhouse gas emissions resulting from:
- Fossil-fuel burning (primary impact)
- Industrial aerosols (secondary impact) (see p.18)

We have seen that on both the global (see p.68) and regional scale (see p.70) climate model simulations which include all of the natural and human factors together can do a fairly good job in reproducing the observed pattern of surface warming. By contrast, simulations that include only the natural factors are unable to reproduce nearly any of the observed surface temperature changes. This observation holds when we compare simulations at the detailed spatial level, as seen in the global maps to the right.

Climate model simulations that include only natural factors reproduce hardly any of the observed surface temperature changes.

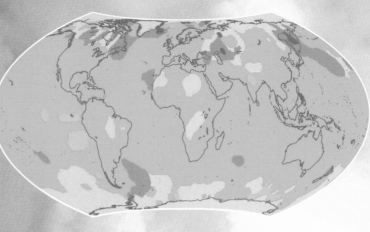

Natural climate model surface temperature calculation 1979–2005

Human and natural climate model surface temperature calculation 1979–2005

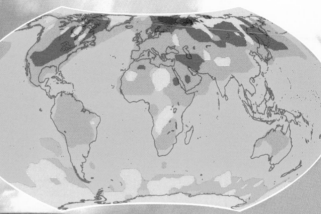

Actual recorded surface temperatures 1979–2005

WARMING PATTERNS

The pattern of warming over the past few decades is not reflected in the models that only take into account the impacts of natural forces alone (top map). Instead, actual observations (bottom map) correspond closely with model predictions that take into account the impacts of both natural and human forces (middle map).

Surface temperature key

-0.6   -0.4   -0.2   0   0.20   0.4   0.60   No Data

TEMPERATURE CHANGE (°C)

Coal fire
Coal-fired power stations are among the primary sources of industrial greenhouse gas emissions.

...other fingerprint is the pattern of ...pected atmospheric temperature ...anges. Different factors, natural and ...man-generated, have different effects ... the different layers of the atmosphere ...e p.38). Increases in solar output ... predicted to warm essentially the ...tire atmosphere from top to bottom. ...lcanoes cool the lower atmosphere (the ...posphere) slightly, and warm the mid-...el atmosphere (the stratosphere).

By contrast, human-generated increases in greenhouse gases are predicted to warm the lower atmosphere (the troposphere) substantially, at the expense of cooling the upper atmosphere (the stratosphere and above).

Human-generated greenhouse gas concentration increases are thought to be the primary cause of atmospheric temperature changes over the past century. Not surprisingly then, the

Solar effect on atmospheric temperature 1890–1999

Volcanic effect on atmospheric temperature 1890–1999

Human-generated greenhouse gas effect on atmospheric temperature 1890–1999

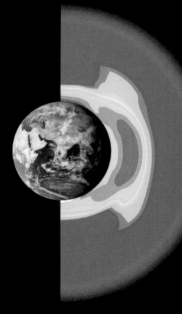

...ATMOSPHERIC TEMPERATURE CHANGE

...The model-predicted pattern of atmospheric ...temperature change taking into account the ...mpacts of both natural and human influences ...far right) looks much like the greenhouse ...gas pattern alone (second from right). This ...s because human-generated increases in ...greenhouse gas concentrations are believed

Atmospheric temperature key

< −1.2 −1 −0.8 −0.6 −0.4 −0.2 0 0.2 0.4 0.6 0.8 1 1.2
TEMPERATURE CHANGE (°C)

Atmospheric layers are not drawn to scale; height has been exaggerated in order to show color variations as clearly as possible.

expected pattern of atmospheric temperature change due to all impacts combined (natural and human) looks much like the greenhouse gas pattern alone. Indeed, we see that this pattern matches the observed pattern of temperature change in recent decades (see p.38).

Combined effect of human and natural forces on atmospheric temperature 1890–1999

Natural fireworks
Stromboli, in Italy, is one of the most active volcanoes on Earth.

Part 2
Climate Change Projections

 Projections of how Earth's climate will change are uncertain. They depend on both the unknown future trajectory of greenhouse emissions, and the uncertain response of the climate to these emissions. Nonetheless, researchers can draw certain conclusions given best-guess scenarios of fossil-fuel burning and the average predictions of theoretical climate models. Scientists can project, for example, that for "middle of the road" emissions scenarios, the globe is likely to warm by several more degrees Celsius by the end of the 21st century. This warming is likely to be associated with a dramatic decrease in Arctic sea-ice, an acceleration of sea level rise, and increased drought, flooding, and extreme weather for many regions of Earth.

How sensitive is the climate?
Modern evidence

To determine the potential magnitude of future global warming, scientists find it useful to estimate something they call "climate sensitivity."

- **Climate sensitivity** defines the amount of warming (in degrees Celsius) that we can expect to occur when there is a change in the factors that control climate. It is a way of placing a numerical value on how much our planet will warm in response to future increases in greenhouse gas emissions. Climate sensitivity is typically expressed in terms of the expected surface warming that will occur in response to a doubling of atmospheric CO_2 levels from their pre-industrial level of roughly 280 parts per million (ppm) by volume.

- **Equilibrium climate sensitivity** takes into account the fact that the full amount of warming in response to an increase in greenhouse gas concentrations may not be realized for many decades, due to sluggish ocean warming. In plain language, this means that if we say equilibrium climate sensitivity is 3°C, we mean that Earth will eventually warm by 3°C if CO_2 levels reach 560 ppm by volume. At current rates of fossil-fuel burning, this doubling of CO_2 levels is expected to occur mid-way through the 21st century. However, the resulting warming may not fully be experienced until at least 2100.

So how do scientists estimate climate sensitivity?

Using climate models, scientists compare observed temperature changes (from the instrumental record of the past 150 years) with simulations of temperature changes over this same time frame (see p.68). In order to determine the actual climate sensitivity, certain types of climate models are tuned to different climate sensitivity values. Scientists then determine which of these sensitivity values best match the observed temperature changes. By looking at the various climate sensitivities that fit reasonably well with the actual temperature record, the scientists can quantify the uncertainty of the estimated climate sensitivity values. This range is quite large if only the relatively short instrumental record is used.

SIMULATED VS ACTUAL SURFACE TEMPERATURE CHANGES

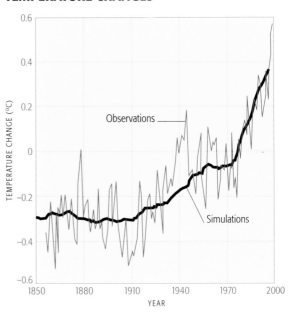

A climate sensitivity of 2.7°C was assumed in these simulations, which took into account both human and natural factors over the 150 years for which surface temperature observations exist.

Deep ocean temperature

Over the past few decades, it has also been possible to make use of temperature measurements from the deep oceans. While such data are useful, they are limited by an even shorter available record.

SIMULATED VS ACTUAL DEEP-OCEAN TEMPERATURE CHANGES

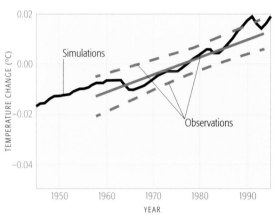

A climate sensitivity of 2.9°C was assumed in these simulations, which took into account both human and natural factors over the 50 years for which deep-ocean temperature observations exist.

Do we need more data?

Overall, these simulations yield an estimate of equilibrium climate sensitivity of roughly 3°C. In other words, it is estimated that Earth's surface will warm by 3°C if CO_2 concentrations double. Uncertainties, however, are large.

Modern instrumental observations could be consistent with a climate sensitivity anywhere from 1.5°C to 9°C.

With such a wide temperature range, the effect of climate change could be anything from essentially negligible to wholly catastrophic. This uncertainty is inevitable when we only have access to a short (150 years or less) record. There are so many different natural and human factors that are simultaneously at play, and each has impacts that are individually uncertain. For this reason, scientists turn to other longer-term sources of information. In the following pages, we show how this is done.

Floating monitors
Weather buoys gather data to keep scientists informed of climate changes on the surface of the oceans.

How sensitive is the climate?
Evidence from past centuries

Another way to estimate climate sensitivity is to study responses to changes in the natural factors governing climate in previous centuries. Using information from climate proxy data (see p.40), such as tree rings and ice cores, scientists estimate how the average temperature of the northern hemisphere varied during the past millennium. They also estimate how the natural factors influencing Earth's climate changed over this time frame, and then compare the two sets of data. Of course, it is important to note that all these estimates come with substantial uncertainties.

Information about how the factors that govern climate have changed takes many forms. Sunspot records are available from the early 17th century up to modern times. Chemical substances that fall to the surface in snow and become trapped in ice cores can be used to track solar activity even further back in time. Explosive volcanic eruptions can be documented through analyses of the aerosol deposits they left behind in ice cores. The long-term increase in greenhouse gas concentrations since the advent of industrialization is documented in the content of air bubbles trapped within the ice (see p.32).

Consistent sensitivity estimates

The equilibrium climate sensitivity estimate of 2–3°C derived from proxy data is similar to estimates produced using the modern record (see p.78) and geological data (see p.82).

Sunspot record
Measurements of sunspots date back to the early 1600s, when they were first recorded with telescopes by European astronomers, such as Galileo. Modern satellite measurements demonstrate that the sun is slightly brighter in years with high sunspot counts.

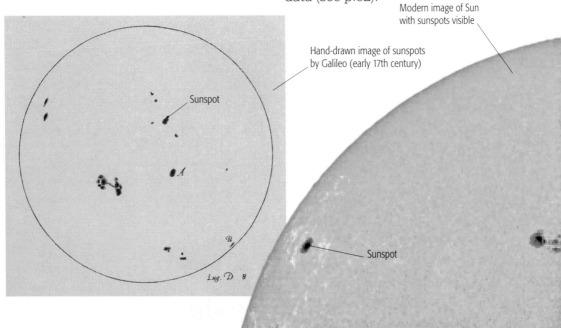

Modern image of Sun with sunspots visible

Hand-drawn image of sunspots by Galileo (early 17th century)

Sunspot

Sunspot

NORTHERN HEMISPHERE TEMPERATURE CHANGES OVER THE PAST SEVEN CENTURIES: SIMULATED VS ESTIMATES FROM PROXY DATA

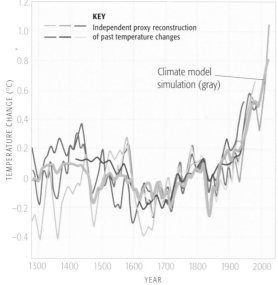

KEY
Independent proxy reconstruction
of past temperature changes

Climate model simulation (gray)

Climate scientists compare model predictions with estimated changes in average temperatures in the northern hemisphere derived from proxy data. The proxy temperature estimates match the model simulations well when the assumed equilibrium climate sensitivity is 2–3°C, meaning that a doubling of atmospheric CO_2 concentrations will lead to a roughly 2–3°C warming of the globe.

ESTIMATES OF NATURAL AND HUMAN IMPACTS ON CLIMATE OVER THE PAST 1000 YEARS

Volcanic impact

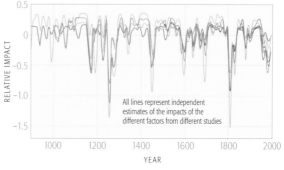

All lines represent independent estimates of the impacts of the different factors from different studies

Solar intensity impact

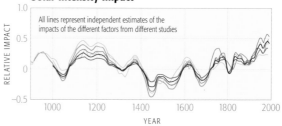

All lines represent independent estimates of the impacts of the different factors from different studies

Human impact

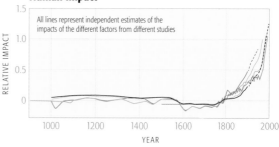

All lines represent independent estimates of the impacts of the different factors from different studies

Scientists drive climate models with the estimated impacts of both natural and human forces. Individual volcanic eruptions have a significant short-term impact, but collectively volcanoes can drive long-term changes when their frequency and magnitude change from one century to the next. Fluctuations in solar output take place on centennial-long timescales. Human-caused greenhouse gases have ramped up dramatically over the past two centuries.

How sensitive is the climate?
Evidence from deep time

As we've seen, studies of climate change over the last few centuries can provide us with reliable estimates of climate sensitivity. These sensitivities correspond to changes in atmospheric CO_2, ranging from the pre-industrial level of 280 ppm to the 2008 value of 386 ppm. While significant, this range doesn't include the known glacial–interglacial variations in atmospheric CO_2 over the last 650,000 years: at the peak of the glacial periods, atmospheric CO_2 dipped to 180 ppm. The range also comes well short of possible future increases in atmospheric CO_2, which are predicted to reach nearly 2000 ppm. So how do we determine how climate will respond to the significantly elevated levels of atmospheric carbon dioxide anticipated for the future? We need to look to the ancient past for clues.

Sea level today

Clues from deep time
Geologists estimate that ancient atmospheres contained as much as 2000 ppm of CO_2, and even more (see p.40). Therefore studies of ancient climates can provide important information on climate sensitivity for much larger CO_2 ranges.

A turbulent past
For the last 2 million years, Earth has been swinging in and out of glacial conditions, driven by subtle changes in Earth's orbit around the Sun that are amplified by feedbacks in the carbon cycle and climate system. Data from ice cores demonstrate that fluctuations in CO_2 and temperature have gone hand in hand for at least the last 400,000 years. Feedback loops in the carbon cycle make the question of whether CO_2 is driving climate changes or vice versa virtually impossible to answer. Nevertheless, computer models only simulate the observed cooling when input with low atmospheric CO_2 levels.

Then and now
To learn more about how climate responds to different CO_2 levels, let's step back in time to the height of the last ice age (the "Last Glacial Maximum," or LGM) 21,000 years ago. With much less CO_2 in the atmosphere, the world was then quite a

During the last ice age, sea level was 120 m lower

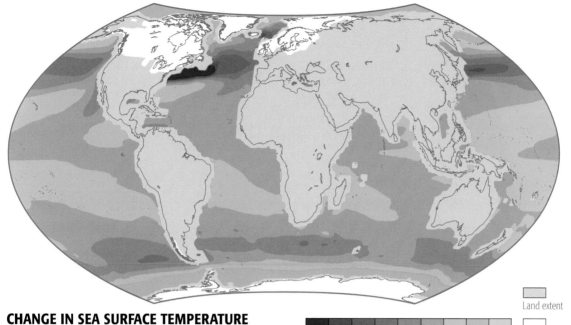

**CHANGE IN SEA SURFACE TEMPERATURE
NOW COMPARED TO 21,000 YEARS AGO**

Sea surface temperature differences between the Last Glacial
Maximum, 21,000 years ago, and today, show that the LGM was
generally cooler. Mid-to-high latitudes experienced more intense
cooling (dark blue), especially near the ice sheets (shown in white).

Land extent

Ice extent

| < | −7 | −6 | −5 | −4 | −3 | −2 | −1 | 0 | 1 |

SEA SURFACE TEMPERATURE CHANGE (°C)
(relative to present temperatures)

different place. The sea level was 120 m
lower, because evaporated seawater had
fallen as snow and formed the vast ice
sheets of the northern hemisphere. A
stroll to the beach from Atlantic City, New
Jersey, would have taken days, since the
shoreline was 80 km east of where it is
today. Based on ice-core gas analyses (see
p.32), we know that the atmosphere's CO_2
content was about 50% of what it is now.
There were a number of other differences
between the LGM and today:

- Atmospheric methane was about one-
  fifth and nitrous oxide was about two-
  thirds of what they are today.

- Vast ice sheets covered much of Canada,
  the northernmost US, Scandinavia, and
  northern Europe. These ice sheets were
  considerably more reflective than the

surfaces they replaced. This accounts
for half of the cooling, since the ice
sheets were reflecting heat rather than
absorbing it.

- Earth's orbital configuration was different
  than it is today (see p.62). Because of this,
  the amount of summer sunshine at high
  northern latitudes was reduced, so snow
  from the winter survived the summer and
  additional ice accumulated.

What can the Last Glacial Maximum teach us about tomorrow's climate?

₂, the LGM, and today

...mate scientists have taken on the
...llenge of assessing the observed climate
...he Last Glacial Maximum. They want
...know if the way climate behaved then
...esponse to changes in CO_2 can help
...understand how it will behave in the
...ure. The compiled data from the previous
...ge indicate that the global average
...mperature at the LGM was 5°C cooler
...n it is today and atmospheric CO_2 levels
...re much lower as well (180 ppm). The
...uilibrium climate sensitivity estimate
...the LGM is 2.3–3.7°C, satisfyingly close
...our other estimates (see p.80). This
...s us that data from the LGM confirm
...predictions for how the climate will
...pond to a doubling of atmospheric CO_2.

Ancient data

How will climate respond to even higher
levels of CO_2 than those experienced in
the glacial–interglacial fluctuations? To
answer that question, we must venture
much further back into Earth's history.
Because ice cores do not go back this far,
there is no direct measure of atmospheric
composition. So geologists have
developed a variety of proxy methods to
study atmospheric CO_2 levels. Each tells
a somewhat different story, but the overall
trends, as shown in the graph below,
are consistent. Atmospheric CO_2 levels
were high 400 million years ago, then fell,
reaching a minimum 300 million years ago.
After that, levels rose and fell, but reached
another maximum about 175 million

CO₂ AND TEMPERATURE IN DEEP TIME

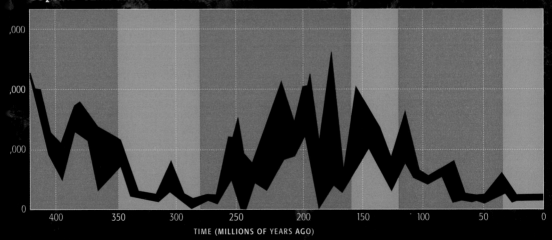

Reconstructions of atmospheric CO_2 levels and global climate
from proxy evidence demonstrate the relationship between the

KEY
Warmer than today

years ago in the late Triassic. They stayed relatively high through the Mesozoic (252–65 million years ago). This was the age of the dinosaurs, when crocodile-like reptiles ventured above the Arctic Circle. Since the late Triassic peak, atmospheric CO_2 levels have generally fallen, reaching another minimum very close to the present day. In the graph below left, you can see that climates (red for warmer than today; blue for as cold, if not colder, than today) generally corresponded to these CO_2 fluctuations. The one exception was the relatively cool interval from 160 to 130 million years ago, which remains unexplained. When combined with the actual paleotemperature estimates, these proxy CO_2 data provide a specific estimate of equilibrium climate sensitivity of 2–5°C for each doubling of atmospheric CO_2, which is entirely consistent with data from the LGM and with the predictions from state-of-the-art climate models. Most importantly, this analysis precludes a weak equilibrium climate sensitivity (less than 1.5°C per doubling). This confirms the general notion that substantial greenhouse warming is the expected consequence of a buildup of atmospheric CO_2.

A billion years of Earth history
Strata in the Grand Canyon contain evidence of large swings in climate that geologists can relate to corresponding fluctuations in atmospheric CO_2 levels.

Fossil-fuel emissions scenarios
Predicting the possibilities

Lurking underneath all our predictions about future climate change is a vexing uncertainty: how will human consumption of fossil fuels and land use practices evolve over the next decades and centuries? The driving forces for consumption are highly complex, involving population growth and per-capita energy demands. These factors are, in turn, closely linked to economic growth and technological advances that can both accelerate consumption and also shift it to other climate-neutral sources.

Possible scenarios

In an attempt to learn more about an uncertain future, climate researchers create and evaluate a range of scenarios for greenhouse gas emissions. This exercise helps them to determine the scope of consequences for a variety of possible future fuel-use scenarios. Initially, these were divided into "business as usual" scenarios, which assume ever-increasing rates of fossil-fuel use, and "conservation" scenarios, which assume some future reduction of use. For the most recent IPCC assessment, experts from around the world developed four basic "storylines", each representing a group

POSSIBLE CO₂ EMISSIONS SENARIOS FOR THE FUTURE

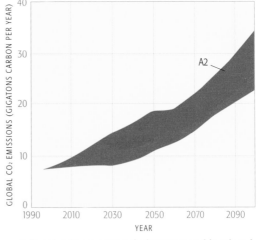

- Substantial reduction of regional differences in per-capita income
- Rapid economic growth
- Peak population mid-21st century, declining thereafter
- Rapid introduction of new, more efficient technology

- Emphasis on national identities and local and regional solutions to environmental protection and social equity issues
- Slow per-capita economic growth and technological advancement
- Continuously increasing world population

of emissions scenarios for the future. The four storylines were designated A1, A2, B1, and B2. To take into account how alternative energy technology developments might affect the climate projections, scientists also refer to three groups of scenarios within the A1 storyline: fossil-fuel intensive (A1FI), non-fossil fuel intensive (A1T), and a scenario that assumes a balanced use of both fossil and non-fossil fuels (A1B). The A1B scenario is a "middle of the road" scenario often used as a basis of comparison in the IPCC report. Each scenario envisions a different future path for Earth and its citizens.

Which one will it be?

Future emissions differ quite dramatically between storylines. The largest growth and cumulative release of CO_2 is associated with the A1FI fossil-fuel-intensive scenarios, while the smallest is associated

with the B1 scenarios. Nevertheless, the B1 scenario leads to an atmospheric CO_2 level stabilized at roughly twice the preindustrial value by the year 2100.

The degree of overlap among these emissions scenarios indicates that very different socioeconomic factors can lead to similar levels of CO_2 emissions in the future. And the spread of values for any given storyline reveals that similar socioeconomic factors can lead to quite different atmospheric CO_2 levels.

The IPCC scientists made no attempt to estimate likelihoods for any of these possible scenarios actually occurring; the uncertainties are simply too large. However, these projections do give climate modelers and social scientists a reasonable range of emissions scenario options with which to work.

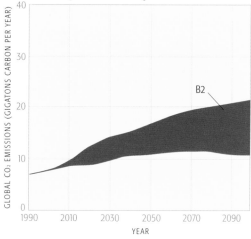

B1 storyline – global utopia

B2 storyline – local utopia

- Emphasis on global solutions to sustainability and environmental protection
- Rapid change to information and service economy
- Peak population mid-21st century, declining thereafter, as in A1
- Reduction in intensity of demand for materials
- Introduction of clean and efficient energy technologies

- Continuously increasing world population, slower growth than in A2
- Intermediate levels of economic development
- Slower development of new energy technologies than B1 and A1
- Emphasis on local and regional rather than global solutions to environmental protection and social equity issues

The next century
How will the climate change?

Scientists use carbon cycle models to convert emissions scenarios (see p.86) into atmospheric CO_2 projections (see figure on left below). These lead to a range of possible trajectories for future climate change (middle figure). The spread of these trajectories is due to differing possible future greenhouse emissions scenarios, and the uncertainty in each trajectory (figure on right) comes from the use of various climate models (see p.64), which differ in their climate sensitivity (see p.78). The results from several different climate models are averaged to yield a single trajectory for each particular emissions scenario.

Temperature changes

The predicted increase in global average temperature from 2000 to 2100 is roughly:

- 1–3°C for the most aggressive emissions scenario (B1 in figures below).
- 1.5–4.5°C for the "middle of the road" scenario (A1B in figures below).
- 2.5–6.5°C for the least aggressive scenario (A1FI in figure below right).

These model projections do not take into account some possible positive feedbacks (see p.94) that could further exacerbate global warming. In certain regions, moreover, warming may be considerably greater than the average predicted for the globe as a whole (see p.92).

It is worthwhile noting that only the most conservation-minded scenarios are likely to avoid warming in excess of 2°C. This is the benchmark rise that is often cited as constituting dangerous human interference with the climate (see p.108).

ESTIMATED CO₂ AND TEMPERATURE TRAJECTORIES FOR VARIOUS EMISSIONS SCENARIOS

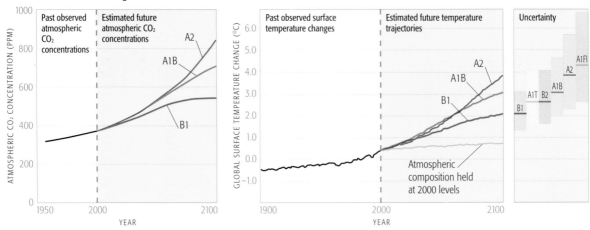

The graph at left shows the estimated CO_2 trajectories for various emissions scenarios. The middle graph shows the resulting temperature trajectories, as predicted by the 23 different state-of-the-art climate model simulations used in the most recent IPCC report. The bold-colored straight lines in the graph at right show the average year 2100 temperatures for various scenarios, while the surrounding lighter bars show the range between the lowest and highest predictions.

HADLEY CIRCULATION PATTERN

In the Hadley circulation pattern, warm moist air tends to rise, cool, and produce rain near the equator. Depleted of its moisture, it eventually sinks as dry air in the subtropics.

Air sinks over the subtropical desert zone

Tropical air flows north in this Hadley cell

Dry desert air flows south

Warm, moist air rises at the intertropical convergence zone, near the Equator

Tropical air carries heat south

Air sinks over the subtropical desert zone

Ferrel cell

Hadley cell

Equator

Hadley cell

Ferrel cell

Precipitation changes

Perhaps of more profound importance than temperature changes are the projected changes in precipitation.

The projected poleward shift in the jet streams of both hemispheres may cause:

- Increased winter precipitation in polar and subpolar regions
- Decreased precipitation in middle latitudes

Poleward expansion of the tropical Hadley circulation pattern will cause:

- Decreased precipitation in the subtropics

A warmer atmosphere will cause:

- Increased precipitation near the equator

PRECIPITATION PROJECTIONS

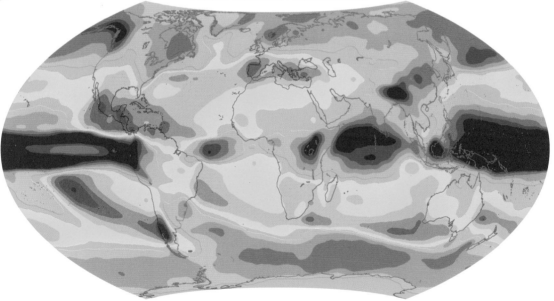

Precipitation pattern changes (relative to 1980–1999) projected to occur by 2100 in response to the so-called "middle of the road" emissions scenario. Note that there is increased precipitation predicted near the equator and in subpolar regions, while subtropical and mid-latitude regions will likely become drier.

< −0.5 −0.4 −0.3 −0.2 −0.1 0 0.1 0.2 0.3 0.4 0.5 >
AVERAGE MODEL-PROJECTED CHANGES IN PRECIPITATION (MM PER DAY) FOR 2080–2099 RELATIVE TO 1980–1999.

More drought, more floods

The combination of decreased summer precipitation and increased evaporation due to warming surface temperatures is predicted to lead to a greater tendency for drought in many regions. The more vigorous cycling of water through the atmosphere favored by a warming globe will lead to greater rates of both evaporation and precipitation. Consequently, more frequent intense rainfall events and flooding can be expected for many regions as well. Other likely impacts of climate change over the next century include increases in extreme weather phenomena (see p.100), and rising sea levels due to melting ice and the warming of the oceans (see p.98).

El Niño Southern Oscillation (ENSO)

The El Niño-Southern Oscillation (ENSO) is a natural irregular oscillation of the climate system, involving inter-related changes in ocean surface temperatures, ocean currents, and winds across the equatorial Pacific. This phenomenon alternates every few years between so-called "El Niño" and "La Niña" events, which influence weather patterns across the globe.

EL NIÑO EVENT

During El Niño events, the so-called "trade winds" in the eastern and central tropical Pacific weaken or even disappear, there is little or no upwelling of cold sub-surface ocean water in the eastern equatorial Pacific, and warm water spreads out over much of the tropical Pacific ocean surface.

LA NIÑA EVENT

During La Niña events, the trade winds in the eastern and central tropical Pacific are stronger than usual, and there is strong upwelling of cold, deep water in the eastern and central equatorial Pacific.

El Niño event

Descending air and high pressure brings warm, dry weather

Southeast trade winds reverse or weaken

Warm water flows eastward, accumulating off South America

Cold upwelling reduced or absent due to weakened trade winds

Low pressure and rising warm, moist air associated with heavy rainfall

La Niña event

Low-pressure system, positioned further to west than normal

Pool of warm water positioned further west than normal

Southeast trade winds

Sea surface cooler than normal in eastern Pacific

Strong upwelling of cold, deep water

LARGE-SCALE IMPACTS OF EL NIÑO (NORTHERN HEMISPHERE WINTER)

El Niño events influence global patterns of temperature and rainfall. The effects of La Niña events are roughly opposite to those shown here for El Niño events

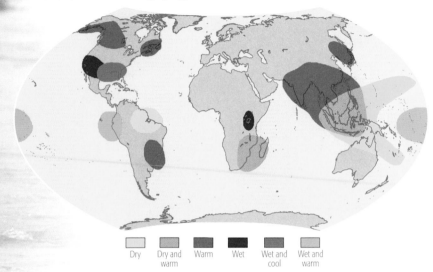

Dry | Dry and warm | Warm | Wet | Wet and cool | Wet and warm

Indonesian rainstorm
A man rides his motorcycle with his son through a flooded street in the city of Tangerang, west of Jakarta. On February 13, 2003, heavy rain pelted Jakarta and surrounding cities for five hours, triggering floods in several parts of the area around the capital.

ENSO VARIABILITY

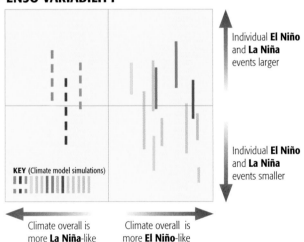

Individual **El Niño** and **La Niña** events larger

KEY (Climate model simulations)

Individual **El Niño** and **La Niña** events smaller

Climate overall is more **La Niña**-like

Climate overall is more **El Niño**-like

Most of the state-of-the-art climate model simulations used in the most recent IPCC report predict a more El Niño-like pattern in response to climate change, but some models predict an opposite, La Niña-like pattern. Models are nearly equally split as to whether the year-to-year ENSO variability is likely to increase or decrease in magnitude.

Uncertain ENSO

Precise regional future climate change predictions are hampered by uncertainties in how global wind patterns and ocean currents will change. Models don't yet agree on the basic question of whether the climate will become more or less El Niño-like in response to human impacts on climate. Since ENSO is such an important influence on regional patterns of precipitation and temperature, such uncertainties translate to an uncertainty about the patterns of regional climate change themselves. If El Niño events become more frequent, then winter precipitation will increase in regions such as the desert southwest of the US, offsetting any trend toward increased drought in the region (see p.48). More El Niño events would also favor a worsening of drought in southern Africa and other regions.

The geographical pattern of future warming

Melting ice field
The extent of summer sea-ice in the Arctic
Ocean has greatly diminished in recent decades.

The pattern of projected warming over the next century is far from uniform. The greatest warming will take place over the polar latitudes of the northern hemisphere, due to the positive feedbacks associated with melting sea-ice. Greater warming is projected for land masses than for ocean surfaces, due mostly to the fact that water tends to warm or cool more slowly than land. Accordingly, there is greater warming in the northern hemisphere, which has a higher proportion of land mass, than the ocean-dominated southern hemisphere. Some of the regional variation in warming is due to changes in wind patterns and ocean currents that are also produced by the changing climate. Relatively little warming, for example, is projected to take place over an area of the North Atlantic ocean just south of Greenland, because weakening ocean currents (see p.60) and shifts in the pattern of the northern hemisphere jet stream favor a greater tendency for cold-air outbreaks in this region.

MODEL-PROJECTED WARMING FOR 2030 AND 2100

These maps shows projected surface temperature changes relative to the average temperatures during the late 20th century for the A1B "middle of the road" emissions scenario (see p.86).

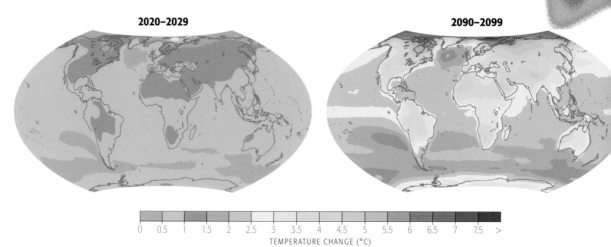

2020–2029

2090–2099

0 0.5 1 1.5 2 2.5 3 3.5 4 4.5 5 5.5 6 6.5 7 7.5 >
TEMPERATURE CHANGE (°C)

Breaking down the projected pattern of warming at continental scales, it is clear that North America is likely to see the greatest warming, while South America and Australia are likely to see more modest warming. It should be kept in mind, however, that precise regional temperature projections are limited by uncertainties in how the El Niño/Southern Oscillation phenomenon and other regional atmospheric circulation patterns will be affected by climate change (see p.90).

CONTINENTAL SURFACE TEMPERATURE ANOMALIES: OBSERVATIONS AND PROJECTIONS

The projected future warming in the "middle of the road" emissions scenario (see p.86) for each continent is well beyond the range of temperature changes seen over the past century.

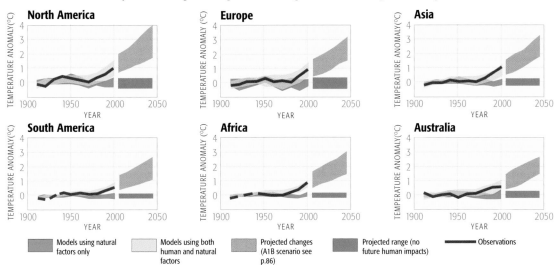

Carbon-cycle feedbacks
Nature's response to CO$_2$

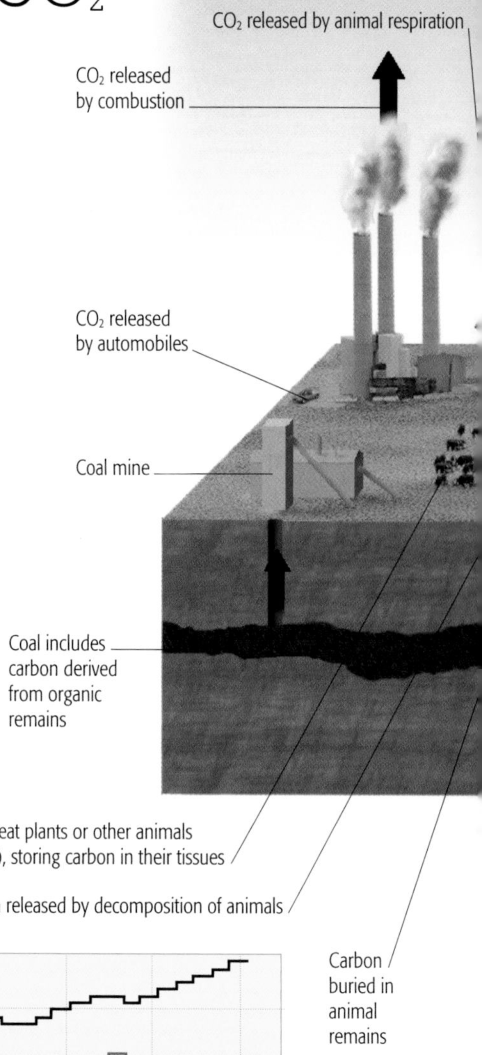

CO$_2$ released by animal respiration

CO$_2$ released by combustion

You might think that all of the carbon dioxide released through fossil-fuel burning and deforestation has simply accumulated in the atmosphere. Yet detailed analysis shows that 45% of the CO$_2$ we've pumped into the atmosphere since 1959 has "disappeared." Actually, scientists know where it went. Much of the "missing" CO$_2$ has dissolved into the ocean, and the rest has been stripped from the atmosphere via photosynthesis and incorporated into living "biomass." (Photosynthesis is the process by which plants, and a few other organisms, use energy from sunlight to convert CO$_2$ into sugar—the "fuel" used by all living things on Earth.) The 55% of the CO$_2$ that has not "disappeared" but accumulated in the atmosphere is termed the "airborne fraction." In effect, nature has already responded to fossil-fuel burning to a certain degree, and somewhat reduced the human impact on atmospheric composition and climate, but nature has its limits and humans are beginning to push up against them.

CO$_2$ released by automobiles

Coal mine

Coal includes carbon derived from organic remains

Animals eat plants or other animals (or both), storing carbon in their tissues

Carbon released by decomposition of animals

Carbon buried in animal remains

WHERE DID ALL THE CO$_2$ GO?

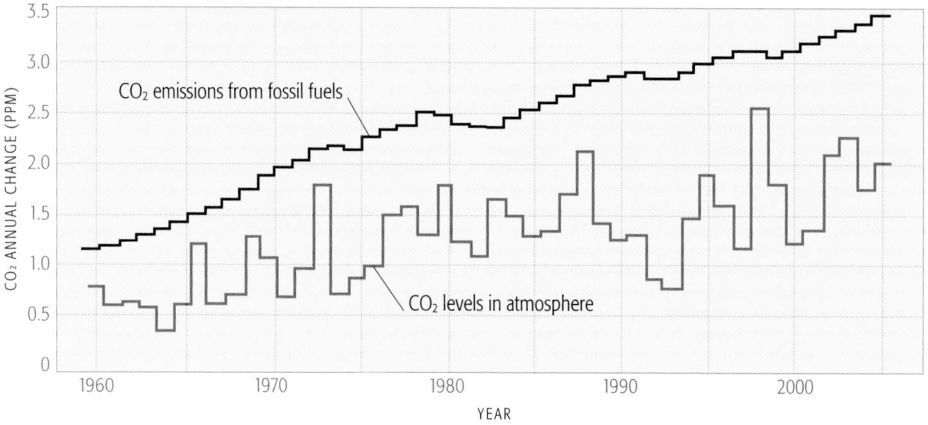

CO$_2$ emissions from fossil fuels

CO$_2$ levels in atmosphere

Y-axis: CO$_2$ ANNUAL CHANGE (PPM) — 0, 0.5, 1.0, 1.5, 2.0, 2.5, 3.0, 3.5
X-axis: YEAR — 1960, 1970, 1980, 1990, 2000

Note that the atmospheric increase does not reflect total emissions; some CO$_2$ has been has dissolved into the ocean or stripped from the atmosphere during photosynthesis.

THE CARBON CYCLE

The main reservoirs of carbon are the atmosphere, the ocean, and vegetation, soils, and detritus on land. Marine life represents a very small carbon reservoir. On multi-millennial time scales, geologic reservoirs also become important. Various processes transfer carbon between these reservoirs, including photosynthesis and respiration, ocean–atmosphere gas exchange, and ocean mixing. The red arrows show the pathway of fossil-fuel CO_2 from its release to its uptake by vegetation or the ocean.

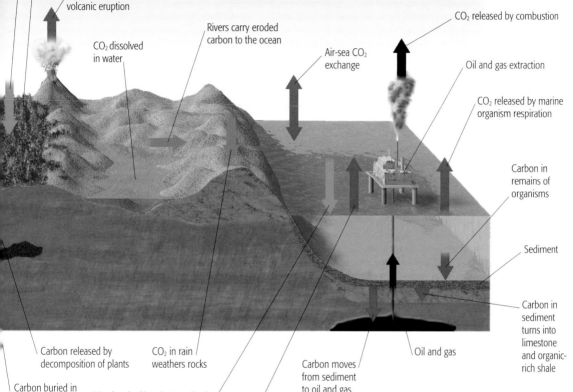

CO_2 released by plant respiration

CO_2 absorbed by photosynthesis

Carbon stored in plant tissues

CO_2 released by volcanic eruption

CO_2 dissolved in water

Rivers carry eroded carbon to the ocean

Air-sea CO_2 exchange

CO_2 released by combustion

Oil and gas extraction

CO_2 released by marine organism respiration

Carbon in remains of organisms

Sediment

Carbon in sediment turns into limestone and organic-rich shale

Carbon released by decomposition of plants

CO_2 in rain weathers rocks

Carbon moves from sediment to oil and gas

Oil and gas

Carbon buried in plant remains

CO_2 absorbed by photosynthesis

Carbon released by marine organism decomposition

KEY

➤ Carbon movement
➤ Photosynthesis
➤ Weathering and erosion
➤ Human carbon transformation

Phytoplankton

Marine phytoplankton like these diatoms play an important role in transferring fossil-fuel carbon dioxide from the atmosphere into the ocean.

95

Positive feedbacks prevail

Despite nature's best efforts to counter our impact on the planet, atmospheric carbon dioxide levels have nonetheless continued to rise, and the planet is warming. Unfortunately, warming reduces nature's ability to absorb carbon dioxide. A number of feedback loops (see p.24) are involved, both positive (enhancing warming) and negative (reducing warming), but positive feedbacks prevail on all but multi-millennial timescales.

Warmer land

Positive feedback:
Soil microorganisms increase their growth and respiration rates as their environment warms. One of the waste products of their metabolism is CO_2. As a result, carbon in soils is now being converted to CO_2 at increasing rates.

Negative feedback:
This release of CO_2 to the atmosphere by soil microorganisms offsets gains made by plants responding favorably in their growth to elevated CO_2 levels (so-called "CO_2 fertilization"; see p.105).

Warmer ocean

Positive feedback:
A warmer ocean has less ability to absorb carbon dioxide, just as an opened can of warm soft drink loses its carbonation and goes flat.

Ocean acidification

Negative feedback:
Acidification of the surface ocean (see p.114) reduces the production of calcium carbonate (limestone/ $CaCO_3$) by organisms such as corals and tiny plankton. When these organisms grow their $CaCO_3$ skeletons (i.e., to calcify), CO_2 is released to the water. So calcification reduces the ocean's ability to take up fossil-fuel CO_2. However, it is predicted that some calcifying organisms will become extinct this century. If calcifying plankton and corals become less abundant, then the resulting reduction in the rate of calcification will slightly increase the ocean's ability to take up carbon dioxide

Goings-on down under
Earthworms, bacteria, and fungi consume plant matter buried in soil, releasing CO_2 at diffuses into the atmosphere above

Soil microorganisms
Bacteria, such as these shown here,
are microorganisms that decompose
organic matter in soils.

Pump problems
Positive feedback:
Calcium carbonate is a
relatively dense mineral
that acts as ballast once
an organism dies, carrying
its decaying tissue to great
depths in the ocean. This
"pump" of carbon removes
CO_2 from surface waters,
allowing more fossil-fuel
CO_2 to be absorbed. Loss
of the ballast via ocean
acidification reduces the
ocean's ability to take up
atmospheric CO_2.

A sluggish ocean
Positive feedback:
A slowing of ocean
circulation in response to
global warming reduces
the mixing up of nutrients
at the ocean's surface,
which slows biological
productivity. This weakens
the action of the biological
pump, further reducing
the ocean's ability to
absorb CO_2.

Rock weathering
Negative feedback:
Increased temperatures
and rainfall stimulate the
weathering of rocks on
land (the process that turns
rock into soil and dissolved
salts in rivers). Atmospheric
carbon dioxide dissolved
in rain forms carbonic
acid, which aids the
rock-weathering process.
Increased weathering,
therefore, removes CO_2
from the atmosphere.

Conclusion
Models that simulate both the carbon cycle and climate have been run with some,
but not all, of these feedbacks taken into consideration. The overall effect of these
feedbacks is a more rapid buildup of atmospheric CO_2, and a warmer climate. For
example, adding carbon-cycle feedbacks to the A2 fossil-fuel emission scenario (see
p.86) leads to an additional 20–220 ppm CO_2 by 2100, and an additional warming
of more than 1°C. This additional warming is reflective of a carbon cycle that is
approaching the limit of its ability to absorb fossil-fuel emissions.

Melting ice and rising sea level

Sea level is predicted to rise with global warming for two reasons. First, water, like most liquids, expands as it warms. A small rise of 0.1–0.4 m is predicted by 2100, depending on the emissions scenario, due to this effect alone. Second, melting ice is likely to have a major impact on the sea level. It is important to note, however, that not all ice plays an equal role here. The disappearance of high-latitude sea-ice (see p.138), while significant in its own right, will not be a contributor. Much as melting ice cubes in a glass of water do not cause the level of water to rise, melting sea-ice will not cause the sea level to rise. On the other hand, melting continental ice will definitely contribute to a sea level rise.

Significant rise

The continental ice resides in two basic forms. First, there are the permanent ice caps and glaciers in mountain ranges at high latitudes, and even at equatorial latitudes (see p.58). Melting all of this ice, however, would only add at most a sea level rise of about 0.5 m.

More significant are the Greenland and Antarctic continental ice sheets. There is evidence that significant melting of the Greenland ice sheet is already underway, but the rate of future melting is difficult to estimate. Model simulations indicate that local warming over Greenland is likely to exceed 3°C by 2100 in the A1B "middle of the road" scenario (see p.92). Current ice-sheet models indicate that such a warming could lead to the eventual irreversible melting of the Greenland ice sheet, resulting in roughly 5–6 m of global sea level rise. Melting of the most unstable part of the Antarctic ice sheet (the West Antarctic ice sheet) could add an additional 5 m. The models suggest that the completion of this melting could take a number of centuries.

Even faster than we thought

But even state-of-the-art models do not account for some newly observed effects that scientists now believe could significantly accelerate the rate of melting. For example, recently it has been discovered that crevices (called "moulins") are forming in melting continental ice. These moulins allow surface meltwater to penetrate deep into the ice sheet and lubricate the base, allowing large pieces of ice to slide quickly into the ocean. If this phenomenon becomes increasingly widespread, it could lead to a far more rapid disintegration of the ice sheets than predicted by any current models.

The sea level is currently projected to rise between 0.5 m and 1.2 m by 2100.

GREENLAND'S MELTING CONTINENTAL ICE SHEET

This map shows changes in the extent of the region of summer melting in Greenland.

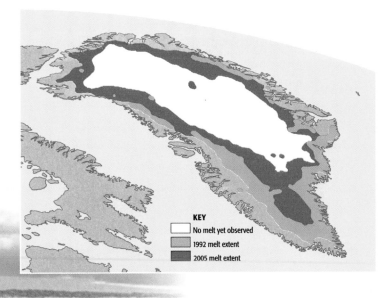

KEY
No melt yet observed
1992 melt extent
2005 melt extent

Moulin in Greenland
Crevices (moulins), such as this one in Greenland, allow surface meltwater to penetrate deep into ice sheets.

PROJECTED SEA LEVEL RISE

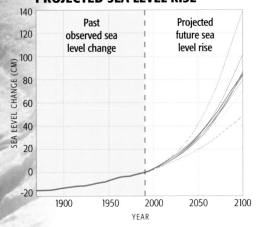

Past observed sea level change

Projected future sea level rise

SEA LEVEL CHANGE (CM)

YEAR

By relying on observations of the past relationships between global temperature and rates of sea level rise, and global temperature projections for the next 100 years (see p.88), scientists can make projections about future sea level rise. The predicted rise by 2100 is between 0.5 m and 1.2 m, depending on the prevailing emissions scenario (see p.86).

Future changes in extreme weather

Frosted roofs
Familiar cold-weather scenes, like these abandoned farmhouses covered with frost near Maupin, Oregon, will become less common as our climate warms.

It is likely that as climate changes, the frequency and intensity of extreme weather events will change. For certain extreme weather events, such as severe frosts and extended heat waves, the science is fairly definitive, and the predicted changes are fairly intuitive. The greater the amount of warming, the more pronounced these trends will be. Changes are predicted to vary regionally.

New trends can be expected to emerge for various types of extreme weather, including heat waves, heavy downpours, and frosts. In the maps below and on the following pages, the colors represent a relative scale. Variations within color fields indicate regions where climate models predict a statistically significant increase or decrease in the quantity in question.

Fewer frosty days

■ As temperatures warm, the probability of frosts (nights when temperatures dip below freezing) will decrease markedly.

■ The greatest decrease in frost days is likely to occur in regions such as interior North America and Asia, where winter temperatures have been traditionally the coldest.

CHANGES IN FROST DAY FREQUENCY

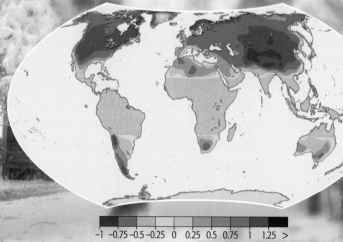

-1 -0.75 -0.5 -0.25 0 0.25 0.5 0.75 1 1.25 >

This map shows projected changes in the occurrence of frost days by the late 21st century (2080–2099) relative to the observed frequency of occurrence in recent decades (1980–1999). A relative scale is used, where a single unit ("1") represents the typical range of year-to-year variations. The term "frost days" refers to the number of days in the year when the minimum nightly temperature drops below freezing.

More heat waves

▪ Heat waves (very high temperatures sustained over a number of days) are likely to become more intense, more frequent, and longer-lasting.

▪ The greatest increase in heat waves is predicted to occur in areas such as the western US, North Africa, and the Middle East, where feedback loops associated with decreased soil moisture may intensify summer warmth (see p.90).

Dry corn
An increase in the frequency of blistering heat waves associated with climate change may make scenes like this one more common in years to come.

CHANGES IN HEAT WAVE FREQUENCY

0 0.75 1.5 2.25 3 3.75 >

This map shows projected changes in the occurrence of heat waves by the late 21st century (2080–2099) relative to the observed frequency of occurrence in recent decades (1980–1999). A relative scale is used, where a single unit ("1") represents the typical range of year-to-year variations. A "heat wave" is a minimum of five consecutive days when the high temperature is at least 5°C above the average.

Wet days and dry days

Most model simulations also indicate that increases are to be expected in the frequency of very intense rainfall events and corresponding flooding. These changes are due to the more vigorous water cycle that will accompany a warmer climate, with greater rates of evaporation from a warmer ocean (see p.24) balanced by more intense precipitation events. Seemingly paradoxical, while many regions are likely to become drier, it is predicted that even in those regions individual rainfall events will become more intense, although longer dry spells will separate them.

Roadway under water

Here a motorist contemplates driving through a flooded road in Beaumont, Texas, following Hurricane Rita on September 24, 2005. Heavy rainfalls will become more common as the atmosphere

CHANGES IN RAINFALL INTENSITY

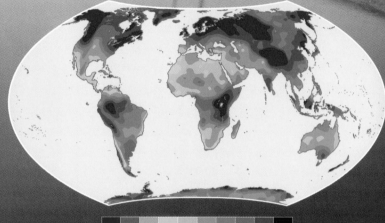

-1 -0.75 -0.5 -0.25 0 0.25 0.5 0.75 1 1.25 >

This map shows projected changes in the pattern of precipitation intensity by the late 21st century (2080–2099) relative to the observed frequency of occurrence in recent decades (1980–1999). A relative scale is used, where a single unit ("1") represents the typical range of year-to-year variations.

CHANGES IN MAXIMUM DRY SPELL LENGTH

-1.25 -1 -0.75 -0.5 -0.25 0 0.25 0.5 0.75 1 1.25

This map shows projected changes in the occurrence of Changes in Maximum Dry Spell Length by the late 21st century (2080–2099) relative to the observed frequency of occurrence in recent decades (1980–1999). A relative scale is used, where a single unit ("1") represents the typical range of year-to-year variations.

Severe storms

It is more difficult to determine how extreme weather events, such as tornados, severe thunderstorms, and hailstorms, will change. This is because such phenomena involve processes that occur at too small a scale to be reproduced in most model simulations. However, it is likely that even if such events do not become more severe or more common in general, individual storms will be associated with more severe downpours and more common flood conditions. This is due to the greater amount of water vapor that a warmer atmosphere can hold. Consequently, this additional water vapor is available to produce rainfall during storm conditions.

Hurricanes and cyclones

But what about the hurricanes and tropical cyclones themselves? We know that there has been a recent trend towards more intense hurricanes in certain basins, such as the tropical Atlantic basin, and that these trends closely mirror warming ocean surface temperatures (see p.56). Also, warmer oceans, all other things being equal, are likely to fuel more intense tropical cyclones, with stronger sustained winds. Model simulations indicate a likely shift towards the strongest (Category 4 and 5) tropical cyclones over the next century, given projected climate changes. There are some important unanswered questions, however. For example, we know that El Niño events change wind patterns over the tropical Atlantic region in such a way as to create unfavorable conditions for tropical cyclone and hurricane formation. And there is still considerable uncertainty about how El Niño will change in response to climate change (see p.90). Climate scientists are currently working to resolve such open questions.

Gusty tempest
Hurricane Dennis' powerful winds, with gusts up to 260 kmh, tossed debris around Cienfuegos, Cuba, on July 8, 2005. Category 4 storms like Dennis (and even stronger Category 5 storms) may become more frequent as tropical sea surface temperatures warm with climate change.

Stabilizing atmospheric CO$_2$
Is a greenhouse world a better world?

The rapid rise of atmospheric carbon dioxide levels over the last two centuries is a clear indication of society's hunger for cheap energy, a hunger that is insatiable and which is being fed an unhealthy diet of fossil fuels. Recognizing this, scientists have been studying future fossil-fuel-use scenarios with substantially elevated atmospheric CO$_2$ levels. The IPCC Fourth Assessment Report paid specific attention to scenarios that had stabilization targets from one-and-a-half to three-and-a-half times the pre-industrial value of 280 ppm; these values are considerably higher than the 386 ppm average of 2008.

The need to act now

A broad range of CO$_2$ stabilization targets has been studied, ranging from 450 ppm to 1000 ppm. The bottom graph at right shows the gradual climb to these levels over the next three centuries. Some interesting characteristics of these curves emerge:

- The lower the stabilization target, the sooner peak emissions of fossil fuel carbon dioxide must occur. In other words, the lower the level at which we want to stabilize the CO$_2$ levels in the air, the sooner we have to cut back on fossil-fuel use. To stabilize atmospheric CO$_2$ levels at 450 ppm, we would need to reach peak usage before 2020.

- Lower stabilization levels can be achieved only with lower peak emissions; while the 1000 ppm target allows CO$_2$ emission rates to double, the 450 ppm target allows them to increase by only 50% or so.

- All stabilization targets require sharp reductions in CO$_2$ emissions following the peak. Low stabilization targets require that emission rates fall below the current rate within a few decades.

EMISSIONS SCENARIOS FOR CO$_2$ STABILIZATION

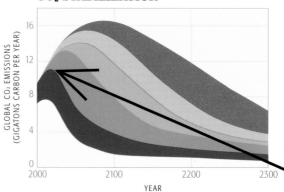

The level at which emissions peak determines the level at which atmospheric CO$_2$ stabilizes.

CO$_2$ STABILIZATION TARGETS

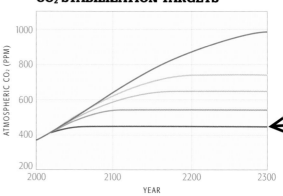

The emission scenarios on this graph show different possible stabilization levels of atmospheric CO$_2$ over the next three centuries. The colors of the lines in this graph correspond to the colors that represent the emissions scenarios in the graph above.

It is also interesting to note that the projected climate changes associated with even the most conservation-minded emission targets are substantial. The long-term warming projected for 450 ppm is about 2°C, an increase that could lead to a sea level rise of half a meter or more (see p.98). In comparison, the "utopian" B1 scenario (p.86-87) stabilizes at an even higher CO_2 level (about 550 ppm) by the year 2100. Moreover, even with the CO_2 level stabilized, the temperature and sea level will continue to rise as the sluggish climate system adjusts to the new atmospheric composition, committing us to continued warming and coastal inundation for decades to come.

Will more CO_2 benefit plants?

Given the serious reductions that would be required to achieve low stabilization targets, some people argue that the higher CO_2 stabilization targets shouldn't be considered failures but rather desirable objectives for beneficial climate modification. This line of reasoning argues that plants require CO_2 for photosynthesis and that more CO_2 should benefit plants (CO_2 fertilization), including the crops that feed the people of the world. Crops grown under ideal conditions in greenhouses with elevated carbon dioxide levels do outperform those grown under ambient atmospheric conditions. In the presence of elevated CO_2, plants do not have to open their pores as wide; this reduces water loss and infection by germs, and encourages growth. However, these benefits cannot be fully realized in nature if other factors, such as a lack of nutrients or inadequate soil, are limiting growth.

The existence of growth-limiting factors, combined with other negative impacts from elevated CO_2 levels—such as ocean acidification and loss of coral reefs (see p.114)—suggests that the "greening of planet Earth" may not be an achievable or desirable outcome of fossil-fuel burning.

To stabilize atmospheric CO_2 levels at 450 ppm, fossil fuel use needs to peak by 2020

With atmospheric CO_2 levels at 450 ppm, global temperature increases by about 2°C and sea level rises by half a meter or more

Part 3
The Impacts of Climate Change

Recent studies indicate that the future impacts of climate change are likely to be far more significant than those observed to date. Human societies, natural habitats, and a myriad of animal and plant species will all be affected by changes in temperature and precipitation patterns in the decades ahead. The precise impacts will depend on the rate and amount of warming, and on the adaptive measures taken by society.

The rising impact of global warming

A world under stress

A list of the potential impacts of global warming on humanity and planet Earth makes for sobering reading. These impacts include a greater tendency toward drought in some regions, the widespread extinction of animal species, decreases in global food production, the loss of coastline and coastal wetlands, increased storm damage and flooding in many areas, and a wider spread of infectious disease. Stresses such as these could, in turn, lead to increased competition for natural resources, over-taxed social services and infrastructures, and conflict between regions and nations. Sustainable approaches to development will be necessary to decrease the vulnerability of society, ecosystems, and the environment to future changes.

Dry lake bed
Scenes like this one from Death Valley, California, may become ubiquitous in the southwestern US if persistent widespread drought takes hold in a warmer world.

EFFECT OF FURTHER TEMPERATURE CHANGES

GLOBAL WARMING IMPACT SCALE

+4.6°C

Global economic losses of up to 5% of GDP

At least partial melting of Greenland and West Antarctic ice sheets, resulting in eventual sea-level rises of 5–11 m

+3.6°C

Substantial burden on health services

Decreases in global food production

About 30% of global coastal wetlands lost

+2.6°C

Changes in natural systems cause predominantly negative consequences for biodiversity, water, and food supplies

Millions more flood victims every year

Widespread coral mortality

+1.6°C

Human mortality increases as a result of heat waves, floods, and droughts

9% - 31% species extinction

+0.6°C

Amount of global warming
(°C increase over preindustrial–before 1750's)

Decreases in water availability; more frequent droughts in many regions

• Wildfire risk increases, as do flood and storm damage

• The burden from increased incidence of malnutrition and diarrhoeal, cardio-respiratory, and infectious diseases escalates

2100 2090 2080 2070 2060 2050 2040 2030 2020 2010

109

Is it time to sell that beach house?

Ten percent of the world's population lives in coastal and low-lying regions, where the elevation is within 10 m of sea level. In some places, such as Bangladesh, this population figure is nearer to 50%. Rising sea level (see p.98), increasing destruction associated with tropical storms (see p.56), increasing coastal erosion, and larger wave heights all pose serious threats to coastal and low-lying regions.

Soggy cities

The most obvious threat associated with global sea level rise is coastal inundation. Coastal regions across the globe are potentially at risk. In North America, for example, significant loss of land on the mid-Atlantic and northeast coastlines could occur with just 6 m of sea level rise —an amount of sea level rise that would result from the melting of the Greenland ice sheet (see p.98). Substantial portions of Europe's "low countries"of Belgium and the Netherlands would also be submerged with a sea level rise of 4–8 m.

SOUTHERN FLORIDA

Southern Florida would be submerged if it were to experience between 4 m and 8 m of sea level rise.

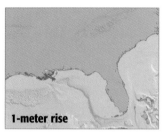

1-meter rise

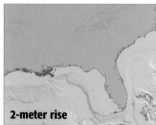

2-meter rise

4-meter rise

NORTHEAST COASTLINE

Most of New York City and Boston could be submerged if sea level were to rise by 6 m.

KEY
Land submerged if sea level rises

6-meter rise

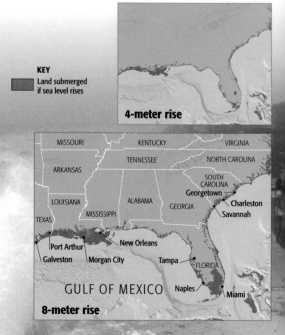

8-meter rise

Human loss

Even those coastal regions not inundated by higher sea levels will be subject to increased exposure to flood and storm damage, more intense coastal surges, and altered patterns of coastal erosion. Associated impacts are likely to include loss of human life, damage to human infrastructure and real estate, degraded water quality, and decreased availability of fresh water due to saltwater intrusion. Coastal habitats will be lost if water levels and wave heights substantially increase. Significant population displacement will also be a factor. Communities, habitats, and economies on all of the major continents will be affected by even just 1 m of sea level rise. The cost rises dramatically at 5 m and 10 m. The melting of the Greenland ice sheet will likely lead to 5–6 m of sea level rise, and the melting of a large part of the West Antarctic ice sheet would probably result in at least an extra 5 m (see p.98).

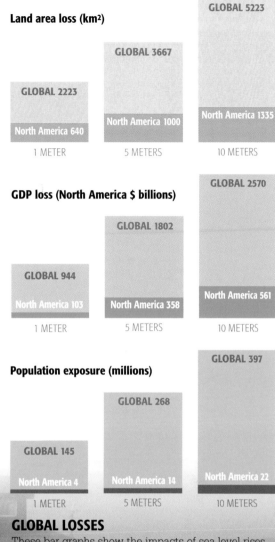

Land area loss (km²)

- GLOBAL 2223 — North America 640 — 1 METER
- GLOBAL 3667 — North America 1000 — 5 METERS
- GLOBAL 5223 — North America 1335 — 10 METERS

GDP loss (North America $ billions)

- GLOBAL 944 — North America 103 — 1 METER
- GLOBAL 1802 — North America 358 — 5 METERS
- GLOBAL 2570 — North America 561 — 10 METERS

Population exposure (millions)

- GLOBAL 145 — North America 4 — 1 METER
- GLOBAL 268 — North America 14 — 5 METERS
- GLOBAL 397 — North America 22 — 10 METERS

GLOBAL LOSSES

These bar graphs show the impacts of sea level rises of 1 m, 5 m, and 10 m. Total global and North American losses are shown. Note that losses are sizeable even in the event of 1 m of sea level rise, which could plausibly occur by the end of this century.

Rough tides
The waves generated by Hurricane Felix encroached on evacuated beach houses on Virginia Beach, Virginia, in August 1995. Homeowners in this community face the dual threat of stronger storms and rising sea level as a result of climate change.

Ecosystems
Worth saving?

Perhaps your drive to work takes you through a reeking, swampy area that's often subject to flooding. Imagine a future in which this lowland quagmire is filled and a new interstate is built, conveying you rapidly through the area with your windows closed and your eyes fixed straight ahead. An improvement? In some ways, perhaps. But was value lost when the ecosystem was destroyed? Is the attractive pond the transportation department built when it reshaped the area serving the same function as the wetland that was destroyed? This begs a further question: are wetlands and other natural ecosystems of any real use to us?

Why we need wetlands

Wetlands provide an important service to their surroundings. Storm waters flowing into a wetland lose energy and spread out across a broad area, thus reducing flooding downstream. Furthermore, sediment and contaminants, such as iron and acidity from mine runoff and nitrogen from farm fertilizers, are removed as water percolates through wetlands before entering our drinking and irrigation water. Wetlands are also biologically diverse ecosystems, providing homes for endangered species and a refuge for migrating birds. This makes them ideal places for hiking, bird-watching, canoeing, and fishing.

What is an ecosystem?

An ecosystem consists of interdependent communities of plants, animals, and microscopic organisms, and their physical environment. All these different elements interact and form a complex whole, with properties that are unique to that particular combination of living and non-living elements. Ecosystem boundaries are generally delineated by climate: desert ecosystems in the subtropics, tropical rainforest ecosystems near the equator, and tundra ecosystems near the poles. As climates have changed in the geologic past, ecosystems have shifted in response. But past climate changes were slower than the projected future changes. Will the ecosystems of today be able to adjust their boundaries as the climate changes, or will they be stranded with incompatible climates? And why, for that matter, should it concern us?

Worth saving

Ecosystems are valuable to humanity because they assist us with:

- **Provisions:** food (seeds, fruits, game, spices); fiber (wood, textiles); medicinal and cosmetic products (dyes, scents)

- **Environmental regulation:** climate and water regulation; water and air purification; carbon sequestration; protection from natural disasters, disease, and pests

- **Cultural benefits:** appreciation of, and communion with, the natural world; recreational activities

Ecosystems are reasonably resilient to change, including a modest amount of human disturbance. But there are limits to this resilience, and human activity is already pushing those limits in some ecosystems (see p.114). Climate change not only challenges the persistence of ecosystems: it may also lead to the extinction of species that cannot adapt or migrate sufficiently rapidly (see p.118).

Unlike local highway construction, the stress placed on ecosystems by climate change is an insidious one—less obvious, but perhaps more permanent. As climate regimes shift poleward, ecosystems will likely follow. However, they may be stopped from doing so by natural factors (such as incompatible soils) and human development (roads, cities, agriculture). The result could be widespread destruction of ecosystems, with attendant loss of benefits to society and a significant reduction in global biodiversity.

Humans meet nature
"Alligator Alley", the highway through the Florida Everglades wetlands ecosystem, was designed with numerous underpasses to minimize environmental impact and threats to the surrounding ecosystems.

Coral reefs Will ocean acidification be their demise?

Coral reefs are among the world's most diverse ecosystems. On a single snorkeling adventure in a healthy reef you can see more species of animals than you can during a lifetime of hiking in mid-latitude forests. Reefs also provide food for hundreds of millions of people, a barrier of defense against the ravages of hurricanes and tsunamis, and a tremendous source of tourism income for nations lucky enough to have them grace their coastlines.

But all is not well with coral reefs, and scenes like the ones depicted on the following pages are becoming all too common. Studies conducted on reefs throughout the world document widespread reductions in coral coverage. The National Oceanic and Atmospheric Administration (NOAA) estimates that 10% of coral reefs are already damaged beyond recovery, and that 30% are in critical condition and may die within the next 10 to 20 years.

Healthy reefs
Coral reefs are among the most diverse and productive ecosystems on Earth. Unfortunately, healthy reefs such as the one shown here are becoming increasingly rare.

Unless significant measures are taken to reduce the stress on coral reefs from human activities, 60% of the world's coral reefs may die by the year 2050.

Unhealthy reefs
This coral reef is bleached: the corals have lost their symbiotic algae that give them their characteristic color. Coral bleaching is believed to be caused by excessively hot ocean temperatures.

EFFECT OF DISEASE ON CORAL HEAD

1988

1998

Black band disease, a bacterial infection linked to warmer water temperatures, destroyed this massive coral head in the Florida Keys. The bacteria are concentrated in the black band, which migrates outward over time, leaving dead coral behind.

Causes of decline

The causes of coral reef decline are many, and include:

Natural stressors

- Disease
- Predation
- Out-competition by algae

Human activities

- Overfishing
- Pollutant runoff from land
- Careless snorkelers walking on delicate coral
- Fuel leaks
- Fuel and wastewater discharge from boats, and oil spills

Additional natural factors are being exacerbated by human activity. For example, coral bleaching—the loss of the algae that live in a symbiotic relationship with the coral animal and give it its color—has been directly linked to intervals of exceptionally hot ocean temperatures. Human-induced global warming is likely contributing to this problem.

Marine protected areas

Marine protected areas (MPAs) are being established around the world, and have proven to be effective at staving off coral and fish losses. They have been shown to be of great economic benefit as well. The United Nations Environmental Program estimates that MPAs cost less than US $1000 per square kilometer, whereas the economic value of coral reefs has been estimated at US $100,000–$600,000 per square kilometer.

Unfortunately, MPAs cannot protect coral reefs from warming, nor can they protect corals from the direct effect of increases in atmospheric carbon dioxide. The ocean has absorbed approximately half of the carbon dioxide released by fossil-fuel burning and deforestation (see p.94). When carbon dioxide dissolves in water, it is transformed into carbonic acid, which makes the water less conducive to coral growth. Sadly, a recent study has shown that every square kilometer of the ocean, with the possible exception of a few remote Arctic areas,

Experimental studies suggest that if fossil-fuel burning rates continue to increase, corals will be unable to grow skeletons by the end of this century.

has suffered detrimental consequences from human activities similar to those experienced by coral reefs.

The only way to prevent this catastrophe is to reduce or eliminate CO_2 emissions, or to develop effective means of sequestering CO_2 before it escapes into the atmosphere.

Coral or cars?

We face a stark choice: either we continue emitting carbon dioxide at ever-increasing rates and accept the demise of coral reefs, or save the coral reefs by reducing or eliminating fossil-fuel CO_2 emissions. It's a genuine dilemma, but for those who value the beauty and understand the importance of coral reefs to the global environment and to society, the path is clear.

And it's not just coral reefs that are at stake. A common theme of this book is that for many reasons it may well be in our own best interests to lessen the buildup of

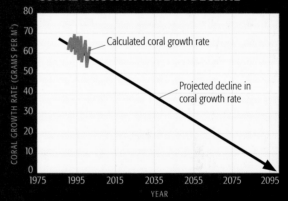

CORAL GROWTH RATE IN DECLINE

Calculated coral growth rate

Projected decline in coral growth rate

This graph shows the projected decline in coral skeleton growth rate (calcification) through this century. The data shown in red represent a calculation of coral growth rate (in grams of skeleton produced per square meter of coral per year) based on observed changes in CO_2 buildup in the ocean near Hawaii. The black line shows the extrapolated trend for an A1B "middle of the road" scenario (see p.86). If greenhouse gas emissions continue to escalate, corals won't be able to grow by the year 2100 or so.

The highway to extinction?

The diversity of species on planet Earth today is the result of millions of years of evolutionary interaction between life and its environment. Human intervention is a new, powerful force, which some liken to the forces that led to mass extinctions of life in the past, such as the asteroid impact that probably precipitated the demise of the dinosaurs 65 million years ago.

Polar bears in danger

A case in point is the precarious future of the polar bear, which depends on expansive sea-ice cover to reach and feed on seals. The earlier spring break-up and retreat of the sea-ice is now forcing polar bears to remain on the tundra, where they must fast and survive on reserves of fat. This puts particular stress on female polar bears, which spend the winter in nursing dens and need easy access to seals in spring to rebuild their fat reserves.

Temperature changes and limited water availability can stress individual organisms, which find themselves suddenly outside of their climate "comfort zone." Typically it is

Bears on ice
Since polar bears hunt on sea-ice, the melting of the Arctic ice cap is making it increasingly difficult for the bears to find sufficient food.

not just one species that is affected. A new report indicates that warmer temperatures are translating into lower fish populations off Antarctica. This, in turn, is resulting in lower survival rates for king penguins, threatening a potential population collapse over coming decades. If global warming means that penguins are not getting enough food, then the conditions for the organisms below them in the food chain are probably even worse.

Extinct amphibians

Also of concern is the worldwide loss of amphibians. In a recent assessment, 122 species of amphibians were listed as "possibly extinct" and another 305 as "critically endangered." The golden toad was last seen in the cloud forests of Costa Rica in 1988. In cloud forest ecosystems, mist from clouds is the primary source of moisture. As the climate warms, the trade winds rising up the mountain slope condense at higher elevations, so the clouds shift upwards. This leaves the forests less cloudy and drier, with warmer nights. Birds, reptiles, and amphibians have all been affected, but the golden toad and the harlequin frog are now believed to be extinct. One theory posits that warmer nights favor the growth of the chytid fungus—a potentially fatal pathogen that grows on amphibian skin.

Adapt or die

Amphibians are the first group of organisms identified as at risk of extinction from global warming. Many more will follow as the planet warms, especially if the rate of warming is rapid. Organisms adapt and ecosystems migrate at rates that sadly may be too slow to prevent ecosystem collapse and the extinction of species.

Gone for good?
The golden toad was last seen in the cloud forests of Costa Rica in 1988.

The IPCC report states with medium confidence that 20–30% of plants and animals will be subject to increased risk of extinction if global temperatures rise to 2°C above the pre-industrial level and perhaps 40–70% of species will be at risk of extinction if temperatures rise by 4°C.

We must remember that extinction is irreversible, and that we are inseparably dependent on the diversity of species harbored by our planet, and the goods and services provided by the ecosystems they support (see p.112).

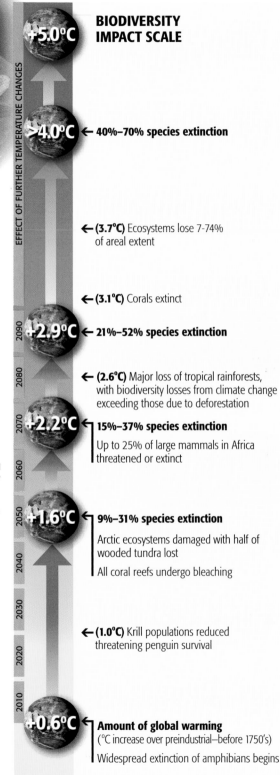

BIODIVERSITY IMPACT SCALE

EFFECT OF FURTHER TEMPERATURE CHANGES

+5.0°C

>4.0°C ← 40%–70% species extinction

← **(3.7°C)** Ecosystems lose 7-74% of areal extent

← **(3.1°C)** Corals extinct

2090 +2.9°C ← 21%–52% species extinction

2080 ← **(2.6°C)** Major loss of tropical rainforests, with biodiversity losses from climate change exceeding those due to deforestation

2070 +2.2°C ← **15%–37% species extinction**
Up to 25% of large mammals in Africa threatened or extinct

2060

2050 +1.6°C ← **9%–31% species extinction**
Arctic ecosystems damaged with half of wooded tundra lost
All coral reefs undergo bleaching

2040

2030

2020 ← **(1.0°C)** Krill populations reduced threatening penguin survival

2010

+0.6°C ← **Amount of global warming**
(°C increase over preindustrial–before 1750's)
Widespread extinction of amphibians begins

Profile James Lovelock

"... for now, the evidence coming in from the watchers around the world brings news of an imminent shift in our climate towards one that could easily be described as Hell: so hot, so deadly that only a handful of the teeming billions now alive will survive."

This most dire of predictions comes from James Lovelock, a British inventor and scientist, in his book *The Revenge of Gaia* (Penguin, 2006). Lovelock is perhaps best known for his "Gaia Hypothesis." According to this hypothesis, all life on Earth, acting like a single organism, participates in planetary-scale regulation of the climate, atmosphere, and oceans. Although controversial, Lovelock has displayed an uncanny ability to originate ideas that evolve from fringe to mainstream.

Working on the NASA Viking mission in the 1960's, Lovelock was responsible for helping NASA determine if life exists on Mars. Believing his colleagues were too

"Earth-centric" in their attempts to determine the existence of extraterrestrial life, Lovelock instead pondered the more general question of what the common characteristics of life might be, and how those characteristics might be expressed in detectable ways on another planet. On Earth, for example, plants produce oxygen and bacteria produce methane (natural gas) in such large amounts that Earth's atmosphere has considerable quantities of both gases, much more than would co-exist on a lifeless planet. It follows that the presence of methane and oxygen in Earth's atmosphere is a signature of a living planet. Lovelock knew that Mars, on the other hand, had a nitrogen–carbon dioxide atmosphere, which is essentially what one would predict of a lifeless planet. Based on this observation, Lovelock asserted that life would not be found on Mars. NASA returned from Mars with important information and fascinating images of our nearest planetary neighbor, but they detected no signs of life. Lovelock's method of planetary life-detection using atmospheric composition is now the central approach of the Terrestrial Planet Finder program at NASA.

Lovelock also conceived of an approach to "terraforming" Mars that would involve melting the ice-caps, which are rich in frozen CO_2, with nuclear warheads. This would release greenhouse gas into the atmosphere and warm the planet to habitable conditions. Unable to publish this idea in scientific journals, Lovelock resorted to writing (with Michael Allaby)

about terraforming in an entertaining science-fiction book, *The Greening of Mars* (Warner Books, 1984). Much later, a NASA-sponsored terraforming study arrived at very similar approaches for making the planet habitable.

The list of Lovelock's prescient ideas extends beyond the scope of this book, as does the societal and scientific impact of his many inventions. For example, he suggested that evergreen trees warm the climates of high latitudes by shedding snow and absorbing more sunlight than snow-covered tundra. This feedback loop is now accepted as a standard component of climate models.

We might do well to carefully consider Lovelock's "outrageous" predictions for the future of humanity in a greenhouse world. In some respects, the dire predictions of the Fourth Assessment Report of the IPCC are more in step with Lovelock's views than with the mainstream thought that prevailed within the scientific community just a few years ago.

Too much and too little
Will floods and droughts really get worse?

There is nothing more precious to living things than water. Changes in the availability of fresh water are of paramount importance in gauging the impacts on society of climate change. These impacts may at first seem contradictory, since increased drought is predicted in many regions, while more frequent intense precipitation events and flooding are predicted for others (see p.88). Such diverse changes result from a complex pattern of shifting rain belts, more vigorous cycling of water in a warmer atmosphere, and increasing evaporation from the surface due to warmer temperatures.

Sunken city
The Mississippi River flooded the city of Kaskaskia, Illinois, in summer 1993. The city, already an island in 1993 thanks to past repeated floodings, never recovered from this episode. Flooding is likely to become more commonplace in many parts of North America and Europe, even in places that are simultaneously suffering from drought.

One of the most significant potential impacts of climate change is diminished or unreliable fresh water supplies.

Globally, water demand is likely to escalate significantly in future decades, primarily due to population growth. Yet this growth is taking place at a time when, in many regions, fresh water resources may be growing more scarce due to climate change.

A combination of warmer water, more intense rainfall events, and longer periods of low river and stream flows will also exacerbate water pollution. Combined with other aggravating factors, such as population growth and increased urbanization, these impacts put intense pressure on fresh water supplies. Since steady running water is required for hydroelectric energy plants, and for cooling towers used in nuclear energy production, decreased river and stream flows can threaten energy resources too.

Slow stream
As a result of severe drought in the Brazilian Amazon, rivers that are the lifeline for local people for transport, supplies and trading are drying up, threatening many communities.

WORLDWIDE
For the more than 15% of the world's population that depends on the seasonal melt of high elevation snow and ice for fresh water, the melting of glaciers and ice caps (see p.98) represents a serious threat.

Serious negative impacts

On balance, the negative impacts of changing precipitation patterns outweigh the benefits. For example, the increases in annual rainfall and runoff in some regions are offset by the negative impacts of increased precipitation variability, including diminished water supply, decreased water quality, and greater flood risks. There is hope, however, that in some cases adaptations (e.g., the expansion of reservoirs) may offset some of the negative impacts of shifting patterns of water availability (see p.150).

FUTURE CLIMATE CHANGE IMPACTS ON WATER

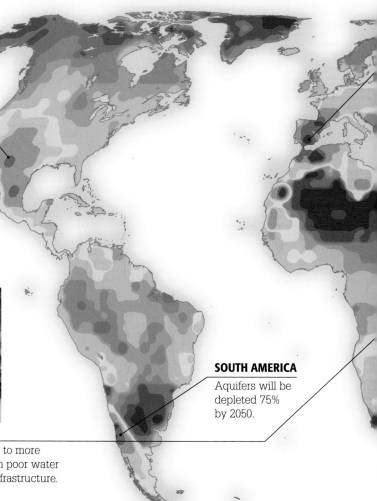

US

A steady increase in the population of cities such as Phoenix and Las Vegas is occurring at precisely the same time that drought conditions are worsening. In the Pacific Northwest, streamflow may have decreased so much by 2020 that the 2007 level of water demand will not be able to be met, and salmon habitat will be lost.

SOUTH AMERICA

Aquifers will be depleted 75% by 2050.

AFRICA

The spread of disease will increase due to more heavy precipitation events in areas with poor water supplies and an overtaxed sanitation infrastructure.

SOUTHERN EUROPE AND THE MEDITERRANEAN

Many arid and semi-arid regions, such as the Mediterranean and parts of southern Europe, southern Africa, and much of Australia, are likely to suffer from increased drought. Electricity production potential at hydropower stations may decrease by more than 25% by 2070.

MORE FREQUENT EXTREME DROUGHT EVENTS

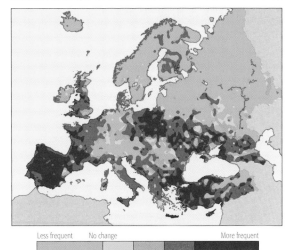

Less frequent   No change   More frequent

<   100   70   40   10   >

SIMULATED RETURN PERIOD (TYPICAL NUMBER OF YEARS BETWEEN CONSECUTIVE DROUGHTS) FOR EXTREME DROUGHT (I.E., DROUGHT WITH A MAGNITUDE EQUAL TO WHAT IS CURRENTLY CONSIDERED A 100-YEAR DROUGHT) BY LATE 21ST CENTURY (2070–2079).

Climate model simulations predict that the spacing between consecutive extreme drought events (defined as once-in-a-hundred-years events) will decrease sharply by the 2070s for the "middle of the road" emissions scenario (see p.86).

BANGLADESH

Areas with increased rainfall and runoff will suffer from an enhanced risk of flooding. The impacts are likely to be especially harsh for regions like Bangladesh, which is already facing the pressures of rising sea level (see p.98).

INDIA

In many coastal regions there will be plenty of water, but it will be the wrong kind! A sea level rise of 0.1 m by 2040–2080 will threaten the fresh water supply.

< -50 -30 -20 -10 -5 0 5 10 20 30 50 >

MEAN CHANGE OF ANNUAL RUNOFF, IN PERCENT, BETWEEN THE PRESENT (1981–2000) AND 2081–2100 (SIMULATED)

Water resources will shift, but not in society's favor.

Is warming from carbon dioxide leading to more air pollution?

Carbon dioxide isn't a pollutant in the typical sense—that is, something introduced into the environment that is a direct threat to human health or to nature. The most notable detrimental health effects of rising CO_2 levels are indirect. They include the negative health effects related to warming we will discuss in more detail later (see p.132), and the intensification of air pollution—a newly discovered phenomenon.

The latest discovery about air pollution utilizes a comprehensive model of climate, pollutant chemistry, and human health effects to calculate the relationships between atmospheric warming and the buildup of pollutants. By directly calculating human health effects, this model is different from other climate models.

Smog is produced when emissions from incomplete fossil-fuel combustion react to produce pollutants. One pollutant of note is ground-level ozone, a lung irritant that also damages crops, buildings, and forests. (Ozone in the upper atmosphere protects us from ultraviolet radiation, but near the ground it acts as a pollutant.) Warming accelerates ozone production and promotes air stagnation, leading to higher ground-level ozone levels.

Los Angeles sunset
Although significant progress has been made to reduce smog in Los Angeles, greenhouse warming may jeopardize future efforts to clean up the air here and in other large polluted cities, such as Houston and Mexico City.

The comprehensive model indicates that for each 1°C increase in temperature there will be an additional 20,000 pollution-related deaths worldwide. This same amount of warming results in even more notable increases in the incidence of asthma and other respiratory illnesses. In the model simulation, there was no question that the cause of these health problems was the buildup of CO_2, because that was the only change to which the model was subjected. If these predictions are correct, society must reckon with yet another unanticipated consequence of fossil-fuel burning: smoggier

In the US, every 1°C temperature rise will result in 1000 extra pollution-related deaths

War

Some climate change skeptics believe that global warming is the sole domain of conservationists, peaceniks, and utopian idealists. Yet policy experts in national security and global conflict, including former CIA directors and White House chiefs of staff, are also worrying about the potential threats posed by fossil-fuel burning.

Why? First, there is the most immediate and obvious reason: reliance on fossil fuels threatens the security of many developed nations by placing them at the mercy of volatile regimes. And there is another, less-discussed security reason: an open Arctic Ocean, forecast to be the norm in "middle of the road" emissions scenarios (see p.138), would have clear international implications. If the forecasts are correct, North American and Eurasian nations will suddenly have new northern coastlines to defend.

But perhaps the most important security concern is the potential for increased competition among nations for diminishing essential resources. As any student of history can tell you, in many parts of the world, such as Latin America, increased stress on resources has historically led to local unrest and unstable regimes.

The possibilities for conflict are countless. For example, the predicted changes in precipitation patterns will naturally create increased competition for available fresh water. Imagine the current Middle East political strife with the added facet of vicious water-resource competition.

Ice floes and flag
This pairing in Glacier Bay National Park, Alaska, eerily foreshadows the emerging connection between melting ice and the threat to US national security.

Environmental refugees

A future with expanded patterns of drought, and conditions unfavorable for agriculture and farming is likely to change people's notions of what constitutes desirable land. Sea level rise and other factors will also potentially make currently inhabited regions inhospitable to humans, thereby increasing competition for habitable land.

The term "environmental refugee" has already been coined to describe individuals fleeing their homelands for more benevolent conditions and climes. Lest one think this merely a hypothetical concept, it should be noted that an estimated 25 million environmental refugees have already been displaced. This is more people than have fled civil war or religious persecution in recent years. Climate change appears to be driving the ongoing migration from the dry Sahel to

OPENING THE NORTHWEST PASSAGE
As the Artic sea-ice retreats and the once-fabled "Northwest Passage" opens up, new sea routes connecting the north Pacific and north Atlantic Oceans are becoming available.

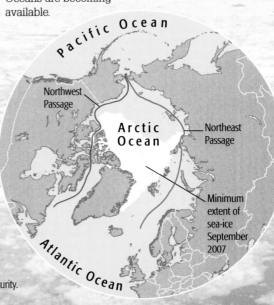

Pacific Ocean

Northwest Passage

Arctic Ocean

Northeast Passage

Minimum extent of sea-ice September 2007

Atlantic Ocean

neighboring regions of West Africa. It also appears to be playing a role in the exodus of people from parts of India, China, Central America, and South Africa.

Ripe for conflict

An optimist might hope that the global threat of climate change will unite the international community as never before, spurring a coordinated campaign among nations to save humanity. Unfortunately, conflict experts foresee the possibility of a different scenario.

A global population predicted to increase to about 9 billion by the mid-21st century, combined with stresses on water, land, and food resources could create the "perfect storm."

As nations around the world exceed their capacity to adapt to climate change, violence and societal destabilization could ensue, leading to unprecedented levels of conflict both between and within nations.

In the "middle of the road" climate change scenario, a combination of worsened drought, oppressively hot tropical temperatures, and rising sea levels could displace a large enough number of people by the mid-21st century to challenge the ability of surrounding nations to accept them, with political and economic turmoil ensuing.

One possible scenario, for example, is that increasingly severe drought in West Africa will generate a mass migration from the highly populous interior of Nigeria to its coastal mega-city, Lagos. Already threatened by rising sea levels, Lagos will be unable to accommodate this massive influx of people. Squabbling over the dwindling oil reserves in the Niger River Delta and the associated potential for corruption will add to the factors contributing to massive social unrest.

Another possible scenario is that drought and decreased river runoff in southwestern North America will strain already water-poor and resource-starved Mexico, leading to increased migration to the US and stress on already delicate diplomatic relations between the two countries.

Even more ominous conflicts can be imagined for the more extreme climate change scenarios. As the nations and peoples of the world compete for diminishing resources, it may become increasingly difficult to establish or maintain stable governance. Indeed, some experts have described worst-case scenarios not so different than those in post-apocalyptic fables such as *Mad Max* and *The Road*.

Famine
More people, less water, less food

Climate change has the potential to seriously undermine the world's food supplies. Sadly, many of the regions most likely to be affected are already finding it difficult to meet existing food demands.

Short-term positives, long-term negatives

Perhaps surprisingly, some regions, such as the US, Canada, and large parts of Europe, stand to benefit at moderate levels of additional warming (1–3°C), thanks to increased crop and livestock productivity.

For these lucky countries, warming will result in longer growing seasons, favorable shifts in rainfall patterns, and higher CO_2 levels, which provide a short-term benefit for plant growth. Indeed, global food production is projected to increase on average with moderate levels (1–2°C) of future warming. But at even those moderate levels, many tropical and subtropical regions—including India and sub-Saharan and tropical Africa, which are already struggling to meet food and pasturing demands—will likely experience a combination of warmer temperatures and decreased rainfall. This will cause a corresponding decrease in productivity of key crops, such as the major cereals.

PROJECTED CLIMATE CHANGE IMPACTS ON CROP AND LIVESTOCK YIELDS

This map illustrates expected changes in crop and livestock yields by 2050 under the "middle of the road" emissions scenario (see p.86). Note that tropical regions suffer more losses as a result of climate change than do temperate regions.

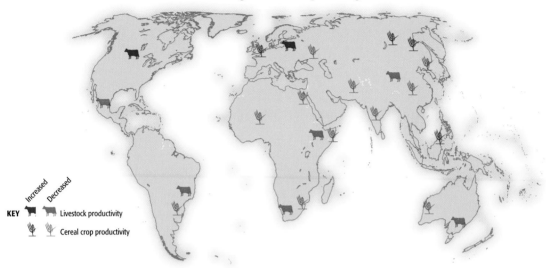

KEY
Increased / Decreased
Livestock productivity
Cereal crop productivity

Moreover, the substantial warming (more than 3°C) predicted to beset us as early as the late 21st century in the "middle of the road" emissions scenario (see p.86) will probably have negative impacts for food crops in all major agricultural regions. Increased forest fires (see p.135) and more frequent disease and pest outbreaks (see p.132) could also diminish available food resources. What's more, the agricultural labor supply is likely to be disrupted by population displacement due to flooding (see p.98) and epidemics.

At sea

Fish populations, and thus commercial and subsistence fishing, may be greatly affected by climate change. The impact of warming is likely to be variable, leading to either increases or decreases in aquatic populations, depending on the location and the species of fish present. Changing ocean circulation patterns also represent a wild card. In particular, the potential weakening in the "conveyor belt" pattern of ocean circulation in the North Atlantic (see p.60) could have significant negative consequences for fisheries in that region.

The good news

Socioeconomic development could partially or even completely offset the negative impacts of climate change on the food supply. There are roughly 820 million people around the globe who are currently undernourished. Taking into account the combined impacts of changing socioeconomic factors and climate change, projections for the number of undernourished people by 2080 range from a reduction to 100 million to an increase to 1300 million.

Somalian famine victims
The line for food in Baidoa, Somalia seems endless. Unfortunately, climate change will only make situations like the devastating famine in Somalia worse.

...Pestilence and death
Human health and infectious disease during times of global climate change

We install heating and air conditioning for indoor climate regulation, build dams for flood control, dig wells to irrigate our fields, and construct dikes to stave off rising seas. But despite our best efforts to insulate ourselves from our natural environment, we remain highly susceptible to climate change, especially when its effects are rapid or unexpected.

Pests and pollen

Of paramount concern is the effect that climate change might have on human health. Diseases can spread as climates change: insects and rodents that carry disease range more widely as climate barriers are lifted. Already there is evidence that vectors such as ticks are spreading to higher latitudes and altitudes in Canada and Sweden. And ragweed is producing more pollen over a longer season as a result of rising temperatures and atmospheric CO_2 concentrations.

Heat can kill

The wake-up call for climate-change-induced human mortality is the European heat wave of 2003 (see p.52). During two extremely hot weeks in early August 2003, nearly 15,000 French people died; across Europe, fatalities approached 35,000. Most of the dead were elderly people who were unable to escape the persistent and oppressive heat. The death rate began climbing several days after temperatures began to rise, and peaked at over 300 additional deaths per day in Paris alone before temperatures finally began to fall. The IPCC report projects an increase in heat-wave incidence with high confidence.

Some like it hot
Mosquitoes and other vectors of disease may spread and flourish as global temperatures increase.

EFFECTS ON HUMAN HEALTH

The IPCC report projects the following climate changes and related health effects in the 21st century.

Predicted climate change (in order of decreasing certainty)	Anticipated effect on human health
On land, fewer cold days and nights	Reduced mortality from cold exposure
More frequent heat waves	Increased mortality from heat, especially among the elderly, infirm, young, and those in remote regions
More frequent floods	Increased deaths, injuries, and skin and respiratory disease incidence
More frequent droughts	Food and water shortages; increased incidence of food- and water-borne disease and malnutrition
More frequent strong tropical cyclones	Increased mortality and injury, risk of food- and water-borne disease, and incidence of post-traumatic stress disorder
More extreme high-sea-level events	Increased death and injury during floods; health repercussions of inland migration

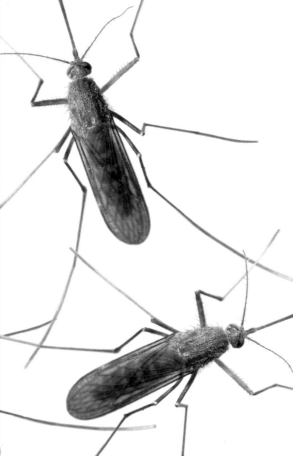

Climate-change-related health impacts will not be uniformly distributed across the world's population. Poor nations will be more susceptible than wealthy ones, because of inadequate access to air conditioning, infrastructure (clean water supplies, electricity, etc.), health care, and emergency response facilities. In all countries, children, the elderly, and the urban poor will suffer disproportionately, as will those people living in low-lying coastal areas.

These threats to human health may serve as a motivating factor for governments to mitigate future climate change. Potential adaptations include raising public awareness, instituting advance warning systems, and improving public health infrastructure in those regions most likely to be hard hit.

Too wet and too hot
Impacts on Europe

The impacts of climate change are already apparent across Europe, from the record-setting heat wave of summer 2003 (see p.52) and the melting of long-standing mountain glaciers, to shifting precipitation patterns and readily observable changes in ecosystems. Some of these changes can be attributed to an atmospheric phenomenon known as the North Atlantic Oscillation. The increasing tendency for the North Atlantic Oscillation to be in its positive phase has favored a stronger, more northerly jet stream over Europe. This may, at least to some extent, be associated with climate change.

Too wet

Average annual rainfall has already increased by nearly 50% over parts of northern Europe. Further increases in winter flooding in coastal regions, and an increased frequency and intensity of flash floods for much of Europe are projected with further warming. Almost 2 million people in the low-lying countries of the Netherlands, Belgium, and Luxembourg could be threatened with flooding in the coming decades as a result of these changes combined with the impacts of rising sea levels.

SELECTED POTENTIAL CLIMATE CHANGE IMPACTS IN EUROPE

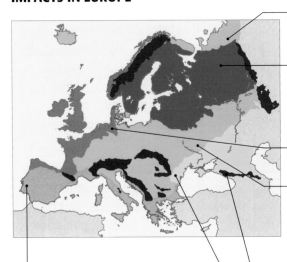

- Thawing of permafrost
- Substantial loss of tundra biome
- More coastal erosion and flooding

- More coastal flooding and erosion
- Greater winter storm risk
- Shorter ski season

- More coastal erosion and flooding
- Stressing of marine ecosystems and habitat loss
- Increased tourism pressure on coasts
- Greater winter storm risk and vulnerability of transport to winds

- Increased frequency and magnitude of winter floods
- Heightened health threat from heat waves

- Severe fires in drained peatland
- Disappearance of glaciers
- Shorter snow-cover period
- Upward shift of tree line
- Severe biodiversity losses
- Shorter ski season
- More frequent rock slides

- More frequent forest fires
- Biodiversity losses escalate
- Negative impact on summer tourism
- Heat wave impacts grow more serious
- Cropland losses as well as losses of lands in estuaries and deltas

- Decreased crop yield
- More soil erosion
- Increased salinity of inland seas

The average rainfall has already increased by 50% over parts of northern Europe.

Too hot

For Europe, the past decade has been the warmest on record for at least the last 500 years (see p.52). Climate model projections indicate that further warming will be concentrated in northern Europe in winter and in southern and central Europe in summer.

It is possible that certain impacts of these changes could be beneficial for human inhabitants. For example, warmer winters could reduce the number of deaths arising from exposure to extreme cold. On balance, however, it appears very probable that the risks to human health will increase. Deadly heat-stress, associated with events such as the 2003 heat wave, will almost certainly become more common in Europe. Increased flooding and warmer winters are also likely to facilitate the spread of water-born, vector-born, and food-born diseases (see p.132).

Across the globe

Already common, heat waves are likely to increase in frequency and intensity over the next few decades as large parts of North America, Asia, Australia, and Africa continue to warm. In the world's cities, urban "heat island" effects, poor air quality, population growth, and an aging population are likely to magnify the impacts of rising temperatures.

Over large parts of North America, Asia, Australia, New Zealand, and South America (e.g., Venezuela and Argentina) increases in heavy rainfall have already led to a higher incidence of floods and landslides. Further increases in such events will likely lead to degraded water quality in these regions (see p.122) and the spread of water-borne diseases.

HUNGARIAN RAINSTORM
This torrential scene took place on Andrassy Street in Budapest, Hungary, in April 2006. More frequent and heavier rainfall can be expected over large parts of Europe as a result of climate change.

The polar meltdown

Slip sliding away
Coastal erosion is a growing problem in the Arctic. The muddy permafrost coastline on the Beaufort Sea of northern Alaska is eroding as waves undercut the melting soil. For scale, the cliff is about 3–4 m high.

Each decade the snow-free period in Arctic Eurasia and North America increases by five or six days, exposing dark ground that absorbs sunlight more effectively than snow. At the same time, the Arctic Ocean's sea-ice cover is decreasing at 7% per decade, exposing the darker sea surface so that the Arctic warms even more. These and other positive feedback loops (see p.24) are amplifying the polar response to the buildup of fossil-fuel CO_2 (see p.92). On average, the Arctic is warming at twice the rate of the globe as a whole, and the region's vast land masses are warming at five times the global average. The situation is less clear in the Antarctic, although a recent study has shown that the Antarctic ice sheet seems to be shrinking for the first time. Continued observation will determine if this really is a downward trend, or simply part of a cycle of shrinkage and growth.

Permafrost thaw

Beneath the Arctic's active layer of soil, which thaws each year, is permafrost—permanently frozen soil. Permafrost is important to human settlement in the Arctic. It has provided a solid and impervious substrate to support building foundations, roadbeds, and pipelines, and to contain sewage ponds and landfill leachate (water that collects contaminants as it trickles through waste).

As a result of recent climate warming and locally released heat from buildings and pipelines, permafrost is now melting. Thawing leads to building collapse, pipeline breakage, roadway degradation, and contamination of surrounding environments. Moreover, methane—a strong greenhouse gas trapped within the permafrost for millennia—is now being released into the atmosphere, further exacerbating warming. By 2050, if current trends continue, the active soil layer will have thickened by 15–50% at the expense of the permafrost below

Meltdown pros and cons

The meltdown may have some benefits to society, such as lower heating costs, greater opportunities for agriculture and forestry, increased river flows (supporting hydroelectric power generation and improved navigation through the Arctic), and easier extraction and transportation of marine resources (including potentially huge hydrocarbon reserves below the Arctic Ocean). However, sea-ice retreat compounded by the rising sea level

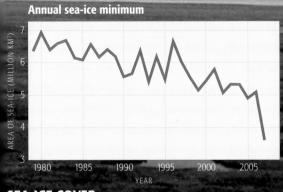

Annual sea-ice minimum

AREA OF SEA-ICE (MILLION KM²)

YEAR

Pacific Ocean

Arctic Ocean

Atlantic Ocean

**Minimum
extent of sea-ice**

September 1980
September 1987
September 1997
September 2007

SEA-ICE COVER

Satellite observations reveal that the seasonal minimum sea-ice cover in the Arctic, which has been declining for decades, fell abruptly in 2007, reaching a record low. Models didn't predict this level to be reached for several decades to come. One possibility is that the models do not take into account important feedback loops that are amplifying the sea-ice response to warming; this raises concerns that our projections for the future are too conservative, and that sea-ice might soon disappear entirely from the Arctic. It is also possible, however, that the 2007 record low was a temporary downward fluctuation, and that there will be a recovery. Climate scientists will be keeping a close eye on what happens over the next few years.

will also lead to coastal erosion and the collapse of roads, buildings, and pipelines. These facilities—and even whole villages and towns— may need to be relocated. Traditionally, indigenous Arctic people lived in small, widespread communities that could easily relocate when necesary. Modern settlement trends are different: now two-thirds of the Arctic population lives in larger settlements of more than 5,000 inhabitants. This concentrated settlement pattern, combined with a greater dependency on infrastructure, reduces the resilience of Arctic populations to environmental change.

Polar politics

The Arctic has garnered much political attention in recent decades; large, untapped reserves of fossil fuels, minerals, and diamonds await better access for exploration and exploitation. Already, the Russian Federation has sent a submersible

to the depths of the Arctic Ocean to plant a miniature flag, and the Danes assert that the geologic feature that traps petroleum in the Arctic Ocean is an extension of Denmark. On a more altruistic note, the developed world is also beginning to recognize and express concerns that the pollution it produces is ending up in the Arctic, with deleterious health and environmental impacts.

The Arctic is warming at twice the rate of the globe as a whole.

Part 4
Vulnerability and Adaptation to Climate Change

 To reduce our vulnerability to climate change, it is necessary that we both adapt to the effects of the existing buildup of atmospheric CO_2 and reduce the amount of CO_2 we are emitting. In the initial stages of climate change, adaptations can help to allay the threat of rising sea level, diminished and shifting fresh water resources, loss of agricultural productivity, and adverse economic effects. Efforts to reduce or mitigate emissions face many obstacles, but ultimately they may be our best hope for the future, when adaptation alone will be insufficient to counter the long-term impacts of climate change on society, the environment, and the economy.

Is global warming the last straw for vulnerable ecosystems?

How human activity has changed the rules of the game

Temperature has been fluctuating on Earth since the planet developed an atmosphere. So, you might say, why is this such a big problem all of a sudden? We know that the increasing ocean temperatures and atmospheric CO_2 levels linked to coral stress and mortality (see p.114) have geological precedence. Carbon dioxide levels have been fluctuating by 100 ppm for the last million years, and climates have changed rapidly as a result, yet corals have survived. Surely, given their past record of adaptation and survival, corals and other organisms will adapt to global warming?

Not necessarily. This "historical precedence" argument has several weaknesses, not the least of which is the fact that fossil-fuel carbon reserves have the capacity to boost atmospheric carbon dioxide levels further and faster than ever before. There also are key differences between the modern adaptation "playing field" and that of the geologic past. Human land use, construction, pollution, and other constraints make adaptation and survival a different game than it used to be. For example, reefs today are under considerable additional stress from human activity such as:

- Wastewater and sediment discharge
- Pesticides
- Ship groundings
- Dynamiting and poisoning for fish collection
- Increasing recreational use
- In the Caribbean, a possible increase in pathogens carried by dust storms that have been exacerbated by human-caused desertification of Africa's Sahel region

These modern factors have increased the sensitivity of ecosystems to global warming, and compromised their ability to "bounce back" from adversity.

In assessing the vulnerability of ecosystems to climate change we must consider:

- The potential rate of change
- Additional stresses imposed by human activity
- Barriers to adaptation and migration imposed by human activity, human settlement, and infrastructure (e.g., roads and pipelines)

Ecosystems in jeopardy

Will we act to reduce the stresses on ecosystems in advance of significant impact? Will we attempt to restore destroyed habitats or create new habitats before changes are irreversible? Are such efforts likely to be successful? There is much that we do not understand about how ecosystems function and what services they provide (see p.112). But many think that if we don't act now, the anticipated stress of climate change will be "the straw that broke the camel's back."

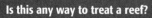

Is this any way to treat a reef?
This wastewater outflow pipe is discharging treated wastewater into the south Florida coastal zone. High nutrient levels in the discharged wastewater promote algal growth that can smother coral reefs.

What is the best course for the coming century?

Up to 1 m of sea level rise could take place by 2100, given "middle of the road" future emissions scenarios (see p.88).

To adapt or to mitigate?

It seems that we have two options: adapt to these changes or mitigate against them. We can take actions to reduce the buildup of carbon dioxide in the atmosphere (mitigation) or to offset the effects of this buildup (adaptation). The problem is that the magnitude of the potential climate change exceeds the capacity for societies and ecosystems to adapt. Moreover, these changes also probably exceed our capacity to mitigate against them, at least for the next few decades (see part 5). Such realities suggest that the best approach is to adopt a plan for the future that is a blend of both adaptation and mitigation, accompanied by technological development to support these efforts and research to guide the technology.

The IPCC report describes the vulnerabilities of countries to climate change with and without mitigation efforts (see maps below and opposite).

Adapt

The vulnerability of each system (i.e., country or region) is related to its adaptive capacity—that is, its "ability or potential to respond successfully to climate variability and change." China and Africa currently

CLIMATE CHANGE VULNERABILITY IN 2100

Vulnerability to climate change can be lessened if mitigation efforts are made and adaptive capacities are enhanced.

KEY (Vulnerability level)

10 Extreme	5 Modest
9 Severe	4 Modest
8 Serious	3 Little
7 Moderate	No Data
6 Moderate	

Assuming current adaptive capacities and no mitigation

With enhanced adaptive capacity and no mitigation

have higher vulnerabilities than developed nations—and will continue to for the next few decades—because of their low adaptive capacity. Even though the adaptive capacities of developing nations are expected to improve with time, overall vulnerabilities are still predicted to remain high. In fact by 2100, even the nations in the developed world, including the US, could be overwhelmed unless steps are taken to enhance adaptive capabilities and to mitigate against the buildup of CO_2 in the atmosphere.

Mitigate

In the example illustrated by the maps below, mitigation means a stabilization of atmospheric CO_2 at 550 ppm (see p.104). This can realistically be acheived only by significant reductions in fossil-fuel burning rates. Decisive action to mitigate against CO_2 buildup leads to a substantial reduction in vulnerability in all but select regions of Africa, China, and Europe. Mitigation efforts primarily benefit developing countries in the short term (over the next few decades). However, by the end of the century, all nations will benefit from taking such actions, as well as from investments in enhancing their adaptive capacity.

Global action required

To improve their adaptive capacity, nations need to incorporate the likely effects of climate change into their strategies for sustainable development and disaster management. Mitigation efforts deal with global factors (atmospheric CO_2), so they will require international agreements (see p.184). The IPCC report makes it clear that the nations of the world need to take aggressive and immediate action to avoid the looming crisis reflected in the maps shown here.

The road not taken
The nations of the world are making decisions now that will make all the difference in how severe the consequences of climate change will be.

Assuming current adaptive capacities and mitigation steps taken to stabilize atmosphere CO_2 levels at 550 ppm

With both enhanced adaptive capacity and mitigation steps taken to stabilize atmosphere CO_2 levels at 550 ppm

It's the economy, stupid!

Clearly we can reduce the potential damage to natural ecosystems by reducing greenhouse gas emissions. What may not be as apparent is that emission reduction can be good for the economy as well. Economists tell us that the formidable cost of emissions reductions may actually be less than the economic damage that will result from climate change. One fairly conservative estimate links unbridled carbon emissions to economic damages amounting to 2–3% of GDP by the year 2100. Society must balance the cost of these damages against the substantive costs of emission reduction.

The cost of carbon

A so-called "integrated assessment model," which takes into account economic considerations as well as climate change, can be used to estimate the "social cost of carbon" or SCC. This is the cost to society of emitting one additional metric ton (tonne) of carbon. The SCC incorporates the climate change and associated economic impacts of carbon emissions over a prescribed time horizon. This time horizon is typically up to the year 2100, but sometimes it also covers the centuries or millennia over which carbon emissions will continue to affect climate. Estimates of the SCC range from a few US dollars per tonne to several hundred. A prominent economist, William Nordhaus, estimates the present SCC at US $30. In other words, the typical American, who drives 10,000 miles per year and thereby emits a tonne of carbon into the atmosphere,

is imposing a cost of US $30 on society. To add insult to injury, this cost impacts society both now and in the future, and the driver is not penalized at all for these damages. Thirty dollars may not seem like much, but think of all the drivers in North America alone, and it really adds up.

> According to Nordhaus, emission reductions over the next several decades could save the global economy US $3 trillion.

Carbon credits

Emissions reduction passes a cost-benefit analysis only when the SCC exceeds the costs of carbon reduction—in the example above, that means as long as it costs less than US $30 to offset the emissions of the typical American car driving 10,000 miles per year. The SCC can be used to set the value of carbon credits—credits issued to nations for reducing carbon emissions, e.g., as part of the Kyoto Protocol—or level of taxation. If the SCC is US $30, a 9 cents per gallon gasoline tax would offset the cost incurred by society for the damage done by driving 10,000 miles a year. Nordhaus equates such a tax with other taxes on harmful practices, such as smoking, and contrasts it to taxes on beneficial activities, such as labor.

potential climate damage over the next several decades

$20,000,000,000

minus the potential reduction

$5,000,000,000

plus the cost of reduction

$2,000,000,000

equals net climate damage

$17,000,000,000

Model limitations

Most of the integrated assessment models do not take into account the distinct possibility of abrupt climate change, which could lead to catastrophic damages without historical precedence. They also do not take into account the possibility that climate change will far exceed current predictions. The resulting damages could lead to astronomical increases in the SCC.

Ethical concerns

Finally, it should be noted that ethical concerns may call for action even when cost-benefit analyses do not. The fact that climate change will probably redistribute resources in a "reverse Robin Hood" fashion (see p.130) is particularly unfair to developing nations, such as Bangladesh. The cost of inaction to these communities may be incalculable.

Given this inherently unbalanced scenario, is it fair for the industrial nations—the primary generators of greenhouse gas emissions—to be the ones calling the shots and determining whether or not action is worth taking?

A finger in the dike

Up to 1 m of sea level rise could take place by 2100, given "middle of the road" future emissions scenarios (see p.98). And sea level might ultimately rise as much as 5–10 m if significant parts of the Greenland and West Antarctic ice sheets melt. Even though that could take several centuries to happen, we may be committed to this eventuality by 2100 if emissions meet or exceed the "middle of the road" scenario estimates. Such a substantial rise in sea level would threaten the viability of coastal settlements worldwide (see p.110).

Even a modest future sea level rise would be problematic, given the the current trend towards the aggressive development of coastlines. This impending crisis remains one of the biggest challenges posed by climate change. Adoption of the most austere stabilization policies could reduce the risks of higher-end projections, but we are already likely committed to moderate (i.e., more than 0.3 m) inundation. Some degree of adaptation will be required in addition to mitigation (see part 5).

Rough water
In February 2007, rough seas and gale force winds caused damage to the promenade between Teignmouth and Holcombe in Devon, UK. Faced with rising sea levels, trying to preserve such structures may prove to be a losing battle.

PROTECTING COMMUNITIES FROM RISING WATER

There are three stages of adaptation that coastal communities threatened with rising sea levels may take.

Protection through engineering

The first stage—the most proactive—seeks to protect the population and infrastructure through engineering solutions (e.g., the construction of "empolderings," which structurally reclaim inundated land, and coastal defenses such as dikes or beach nourishments, which create impediments to inundation).

Accommodating inundation

The second stage of adaptation for coastal communities is accommodation. The schemes employed in this stage allow for some degree of inundation (e.g., the building of flood-proof structures, and the use of floating agricultural systems).

Coastal retreat

The third and final stage of adaptation is retreat. Retreat can take various forms (e.g., managed retreat, the building of temporary seawalls, or the monitoring of coastal threat to determine if and when evacuation is necessary).

The cost of inaction

While some adaptation strategies (e.g., the construction of massive coastal defenses, such as sea walls) could be expensive, the cost of inaction is arguably far greater in terms of lost lives and property. And most cost-accounting doesn't include collateral damages to coastal businesses, social institutions, ecosystems, and the environment. For many island and low-lying regions, adaptations are urgently required; in the absence of adaptation, even the climate changes projected in "middle of the road" scenarios could render these regions unlivable. Indeed one Pacific island (Tuvalu) has already begun to plan for possible future evacuation to New Zealand.

As with other climate change threats (see p.190), ethical considerations arise from the disparity in wealth and resources between developed and developing countries. Adaptation will naturally be more challenging for poorer nations, due to their less-developed adaptive capacity (see p.152), and their limited financial ability to fund costly engineering projects.

water-management strategies

We know that the global demand for fresh water will rise as population grows. We also know that in many regions, increasing demand will coincide with a decreased water supply owing to the impacts of climate change (see p.122).

The challenges ahead

Current water-management practices are unlikely to be adequate for addressing the new and additional challenges resulting from climate change. How will we alleviate both the stress of worsened water pollution on the environment and ecosystems and the increased flood-risk associated with more intense rainfall? How will we address the repercussions of diminished energy resources resulting from reduced river-flow in many regions (see p.124)? And how will we tackle the problem of dwindling drinking and irrigation stores? Fortunately, there are changes in water-management practices that may help us with the daunting challenges ahead.

Adapting management practices

Communities can commit to making "no-regrets" refinements in water-management practices—that is, changes that will be helpful in dealing with the challenges posed by natural year-to-year variations in climate, regardless of whether or not human-caused global warming ultimately proves to be a threat. For example, in the western US, where there is considerable year-to-year fluctuation in drought and flood conditions due to ENSO (see p.90), existing practices designed to deal with this variability could be exploited and refined to accommodate climate change impacts as well. Examples of adaptation procedures include the development of sea-water desalinization facilities, the expansion of reservoirs

Rolling sprinklers
More efficient irrigation methods may prove to be an effective means of adaptation in the face of diminished fresh water supplies.

and rainwater storage facilities, and improvements in water-use efficiency and agricultural irrigation practices.

Planning for the future

Adaptation strategies are already being developed in regions such as North America, Europe, and the Caribbean in recognition of the potential for changes in precipitation patterns and water availability. In some cases, climate model predictions are being taken into account when adaptation procedures are being designed. However, predictions of regional changes in precipitation and drought patterns are still uncertain (see p.89). So are, consequently, the projected changes in river-flow and water levels. As long as such uncertainties persist, it will remain difficult for water managers to develop optimal strategies. Nevertheless, being ready for more of what we have already seen in terms of year-to-year variability makes good sense, no matter what the future holds.

ADAPTATION OPTIONS FOR WATER SUPPLY AND DEMAND

Supply-side	Demand-side
Prospecting and extraction of groundwater	Improvement of water-use efficiency by recycling water
Increase of storage capacity by building reservoirs and dams	Reduction in water demand for irrigation by changing the cropping calendar, crop mix, irrigation method, and area planted
Desalination of sea water	Reduction in water demand for irrigation by importing products
Expansion of rainwater storage	Adoption of indigenous practices for sustainable water use
Removal of invasive, non-native vegetation from river margins	Expanded use of water markets to reallocate water to highly valued areas
Transport of water to regions where needed	Expanded use of economic incentives, including metering and pricing to encourage water conservation

A hard row to hoe

Climate change impacts on agriculture, livestock, and fisheries may jeopardize our ability to provide adequate food for a growing global population (see p.130). Are there adaptations we can make to protect ourselves from these impacts?

Getting ahead of the curve

Some of our agricultural options include changing crop varieties, locations, and planting schedules in response to changing seasonal temperature and precipitation patterns. These techniques, in some cases, reduce negative impacts and, in other cases, even convert impacts from harmful to beneficial.

Adaptive practices could lead to increased crop yields in temperate latitudes, and potentially maintain current yields in tropical latitudes if warming is only moderate. If warming becomes high enough, however, the growing stress on water supplies may increasingly limit the benefits of adaptive strategies.

While adaptations can offset harmful impacts and even yield positive impacts, implementing them will require both a rethinking of governmental policies and new institutions to facilitate changes at the local level. It is therefore important that these measures be integrated into future economic development strategies.

Adaptation measures are not without some cost, both to communities and the environment. And implementation faces some obstacles. As crop yields begin to decrease in response to climate change impacts, there may be greater pressure on farmers to adopt unsustainable practices in an attempt to maximize short-term yields.

Winners and losers

There are ethical considerations that also come into play. Small farmers and subsistence farmers in tropical regions will be most vulnerable to climate change impacts, due to their relative lack of access to institutions that can facilitate adaptation. Yet, their contribution to greenhouse gas emissions is minimal.

Ironically, the farm industry in temperate regions such as the US, which is a major emissions contributor, may stand to benefit slightly from modest warming (see p.130), and is more likely to have access to any needed aid.

Wheat beats the heat
Wheat yields are projected to increase in the extratropical regions like the US, but to decline in tropical regions. Appropriate adaptations could prevent the latter, as long as future warming levels are moderate.

CLIMATE CHANGE IMPACTS ON AGRICULTURE, LIVESTOCK, AND FISHERIES

Sub-sector	Region	Finding	*Alleviation after adaptation*
+3° to +5°C			
Prices and trade	Global	• Reversal of downward trend in wood prices • Agricultural prices: +10% to +40% • Cereal imports of developing countries: +10% to +40%	
Pastures and livestock	Low latitudes	• Strong production-loss in swine and confined cattle	
Food crops	Low latitudes		• *Maize and wheat yields reduced, regardless of adaptation; adaptation maintains rice yield at current levels*
Pastures and livestock	Semi-arid	• Reduction in animal weight and pasture growth; increased frequency of livestock heat-stress and mortality	
+2° to +3°C			
Food crops	Global		• *550 ppm CO_2 (approx. equal to +2°C with no adaptation) increases rice, wheat, and soybean yields by 17%*
Prices	Global	• Agricultural prices: −10% to +20%	
Food crops	Mid- to high-latitudes		• *Adaptation increases all crop yields above current levels*
Fisheries	Temperate	• Positive effect on trout in winter, negative in summer	
Pastures and livestock	Temperate	• Moderate production-loss in swine and cattle	
Livestock		• Increased frequency of livestock heat-stress	
Food crops	Low latitudes		• *Adaptation maintains yields of all crops above current levels; yields drop below current levels for all crops without adaptation*
+1° to +2°C			
Food crops	Mid-to high-latitudes	• Crop growth less likely to be limited by length of growing seasons • No overall change in rice yield; regional variation is high	• *Adaptation of maize and wheat increases yield by 10–15%*
Pastures and livestock	Temperate	• Livestock grazing less likely to be limited by length of growing seasons; seasonal increased frequency of livestock heat-stress	
Food crops	Low lattitudes	• Without adaptation, wheat and maize yields reduced below current levels; rice yield is unchanged	• *Adaptation of maize, wheat, and rice maintains yields at current levels*
Pasture and livestock	Semi-arid	• No increase in productivity of plant growth; seasonal increased frequency of livestock heat-stress	
Prices	Global	• Agricultural prices: −10% to −30%	

Part 5
Solving Global Warming

Adaptation alone is unlikely to avert the most severe impacts of human-caused climate change. Instead, we must take action to mitigate the buildup of atmospheric greenhouse gases that are responsible for observed and projected global warming. Doing so will require that we reduce our reliance on fossil fuels by altering governmental policies and individual lifestyles. Important first steps we must take include forging cooperative relationships with other nations and rethinking how we, as a global community, can satisfy our energy requirements for key sectors of our economy—such as transportation, buildings, and agriculture—with minimal costs to planet Earth.

Solving global warming

There are two ways to mitigate global warming: we can reduce or eliminate fossil-fuel carbon dioxide emissions, or we can remove the carbon dioxide from the atmosphere (see p.178). When we attempt the latter strategy, and remove CO_2, it is referred to as carbon capture and storage (CCS) or carbon sequestration.

Thankfully, there are no insurmountable technological or scientific reasons why we can't employ either strategy: emission reduction or carbon capture and storage. The only barrier is society itself. Although many countries are attempting to reduce them, emissions continue to grow, and atmospheric CO_2 levels are climbing at rates that exceed previous predictions. This is largely because there are few economic incentives for emission reduction, especially in the two largest emitting nations, the United States and China.

Carbon costs

A potential solution to global warming is to translate the social cost of carbon (introduced on p.146) into a carbon cost that is paid by the consumer who emits. A carbon cost is an amount that consumers must pay (as a tax or as part of an emission permit

REDUCTION POTENTIAL AT 3 DIFFERENT CARBON COSTS (US $ per metric ton)

For example, if emission of one metric ton of carbon is taxed at US $20, the forestry sector will potentially reduce emissions by just over 1 Gt CO_2 eq. If the carbon cost or tax is raised to $100, the forestry industry may be incentivized to reduce emissions by more than 4 Gt CO_2 eq. (The preferred unit to measure greenhouse gases is the so-called "CO_2 equivalent," which expresses the combined impact of multiple greenhouse gases in terms of the impact of an equivalent amount of CO_2.)

GT CO_2 EQUIVALENT PER YEAR

Energy supply — $20 $50 $100

Transport — $20 $50 $100

Buildings — $20 $50 $100

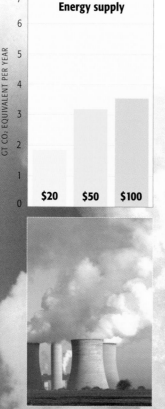

exchange) for the emission of one metric ton (tonne) of CO_2. Taxation, presumably, will not only reduce consumption, but also provide an incentive for the development of non-carbon energy sources.

Emission reduction potential

Mitigation efforts by necessity must span many sectors of the economy, from energy supply, transport, and buildings, to industry, agriculture, forestry, and waste management. The largest potential for emission reductions can be found in some unexpected places, depending on whether we credit reductions at the point of emission or at end-use. If point of emission is credited, then the largest reductions are to be had in the energy supply sector.

However, if we look at end-use, then the buildings sector rises in importance (see figures below). In all sectors, emissions are predicted to decrease as the carbon cost increases. Note, though, that with the exception of the forestry sector, investing larger and larger amounts of money in carbon emission reductions leads to smaller incremental gains—the so-called "law of diminishing returns."

In the pages that follow, we will investigate the ways in which emission reductions might be achieved in each of the major economic sectors, and consider in turn alternative "geoengineering fixes," such as carbon sequestration and the reduction of incoming sunlight.

Where do all those emissions come from?

There is no easy fix for the problem of ever-escalating greenhouse gases. Emissions are traced to all sectors of society and the economy. On the following pages, we will discuss the potential for the mitigation of greenhouse emissions in each economic sector, but first let's examine the bigger picture.

The largest contributor to current global greenhouse emissions is the global energy supply sector. Forestry and industry are the next biggest contributors, followed by transport and agriculture. Most emissions are in the form of CO_2, stemming from fossil-fuel burning (see p.160) and deforestation (see p.174). Methane (CH_4) and, to a lesser extent, nitrous oxide (N_2O), which are primarily associated with agriculture (see p.170), are also significant contributors.

CO_2 equivalent

In order to make comparisons across sectors, it is important to settle on a unit of measurement that takes into account the differing impact of emissions of different types of greenhouse gases. The preferred unit is the so-called "CO_2 equivalent," which expresses the combined impact of multiple greenhouse gases in terms of the impact of an equivalent amount of CO_2. The CO_2 equivalent is typically measured in either megatons (millions of metric tons)

Power lines
More likely than not, wherever you see electrical power lines like these, the electricity they carry was originally generated by the burning of fossil fuels.

or gigatons (billions of metric tons) of CO_2 (abbreviated as Mt/Gt CO_2 eq).

Who emits?

Although the energy supply sector is currently responsible for the largest emissions (nearly 13 Gt CO_2 eq annually), emissions from other sectors have been increasing as rapidly or more so in recent decades. From 1990 to 2004, energy supply emissions increased by roughly a third, while emissions from forestry increased by nearly a half, largely as a consequence of large-scale tropical deforestation. The developed world is currently responsible for the bulk of worldwide greenhouse gas emissions. However, emission rates are increasing most rapidly in the developing world, reminding us that measures aimed at mitigating greenhouse gas emissions must take into account current trends as well as historical patterns.

GREENHOUSE GAS EMISSIONS BY GAS 2004

CO_2 from fossil-fuel burning and deforestation makes up the bulk of greenhouse gas emissions.

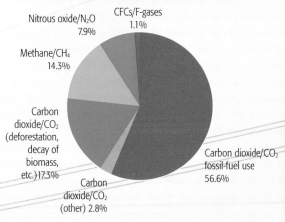

Nitrous oxide/N$_2$O 7.9%

CFCs/F-gases 1.1%

Methane/CH$_4$ 14.3%

Carbon dioxide/CO$_2$ (deforestation, decay of biomass, etc.) 17.3%

Carbon dioxide/CO$_2$ (other) 2.8%

Carbon dioxide/CO$_2$ fossil-fuel use 56.6%

GREENHOUSE GAS EMISSIONS BY SECTOR 2004

The energy supply sector emits the most greenhouse gases.

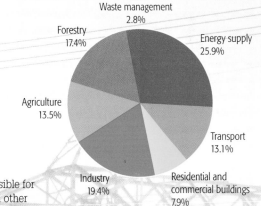

Waste management 2.8%

Forestry 17.4%

Agriculture 13.5%

Energy supply 25.9%

Transport 13.1%

Residential and commercial buildings 7.9%

Industry 19.4%

GREENHOUSE GAS EMISSIONS BY SECTOR IN 1990 AND 2004

While the energy supply sector continues to be responsible for the greatest greenhouse gas emissions, emissions from other sectors, such as forestry, are rising at even faster rates.

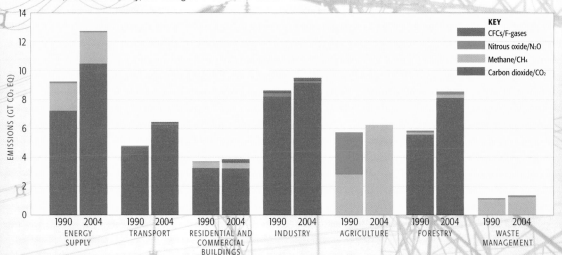

KEY
- CFCs/F-gases
- Nitrous oxide/N$_2$O
- Methane/CH$_4$
- Carbon dioxide/CO$_2$

EMISSIONS (GT CO$_2$ EQ)

| 1990 2004 | 1990 2004 | 1990 2004 | 1990 2004 | 1990 2004 | 1990 2004 | 1990 2004 |
| ENERGY SUPPLY | TRANSPORT | RESIDENTIAL AND COMMERCIAL BUILDINGS | INDUSTRY | AGRICULTURE | FORESTRY | WASTE MANAGEMENT |

Keeping the power turned on

The combustion of fossil fuels, mainly coal and natural gas, generates much of the world's energy supply—the energy we use for electricity generation and heating (oil is primarily used for transport). The energy sector is the single largest source of greenhouse gas emissions, responsible for over a quarter of all worldwide emissions. The primary culprit is CO_2, though methane released during fossil-fuel processing is also significant. Despite recent international efforts to develop and use non-carbon and renewable energy sources, the introduction of new policies, such as carbon trading, and higher energy prices, emissions have increased substantially in recent years. From just 1990 to 2004, annual energy-related emissions increased from roughly 9 to 13 Gt CO_2 eq.

How can we stabilize emissions?

Without widespread governmental action, energy-related emission rates are projected to rise an additional 50% in the coming decades. As emissions continue and their rate increases, stabilizing greenhouse gas concentrations will become ever more challenging. One common misconception is that the "Peak Oil" phenomenon (the projected impending depletion of readily available petroleum reserves) will solve the fossil-fuel emissions dilemma. However, even if oil wells run dry, the primary sources for the energy sector—coal and natural gas reserves—could last for centuries. In reality, meeting the rising global demand for energy supply while simultaneously slowing the rate of fossil-fuel emissions will require a combination of tools. We need to strive for greater efficiency in power generation, an increased use of carbon-free (e.g., nuclear, solar, and wind) or carbon-neutral (e.g., biofuels) energy sources, and the continued development of carbon capture and storage (CCS) technologies.

Energy alternatives

Carbon-free and carbon-neutral energy sources each have their merits and weaknesses. Increased use of nuclear energy (which currently accounts for about 7% of the global energy supply) is limited by a number of factors, including the restricted availability of uranium, security considerations, safety issues, and limited public support. While renewable energy sources such as solar, wind power, and geothermal are currently minor contributors to the global energy supply, government incentives could encourage development and increased efficiency. However, the localized and variable availability of these sources are obstacles to their widespread use in major urban centers. The use of biofuels—such as wood, sugar cane, vegetable oil, and even (yes) dung for heating and cooking—which currently accounts for more than 10% of global energy consumption, could potentially be modestly expanded (see p.172). The increased use of hydropower, another renewable energy source, faces opposition due to the potential environmental threats posed by major damming projects. In many regions, the viability of hydropower may also be threatened by climate change itself, i.e., shifting precipitation patterns (see p.122).

Wind power
Wind turbine generators turn at Scroby Sands off the coast of Norfolk, UK. The wind farm can be seen from Great Yarmouth.

No easy answers

In short, there is no easy way to meet the world's rising energy demands in a climate-friendly manner. All options need to be taken into consideration, at least in the short term. While the developed world has the highest per-capita energy demand, the most rapid growth in energy use is now taking place in developing countries, such as India and China. Efforts to decrease fossil-fuel emissions therefore will require cooperation between the developed and developing world (see p.184).

WORLD PRIMARY ENERGY CONSUMPTION BY FUEL TYPE IN MEGATONS OF OIL EQUIVALENT (Mt OIL EQUIVALENT)

Despite modest increases in the use of renewable energy resources in recent decades, fossil-fuel sources (gas, coal, and oil) continue to supply the lion's share of the world's energy.

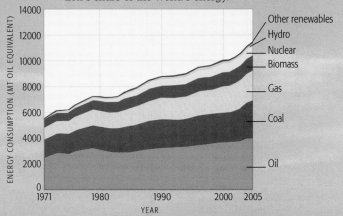

WORLD ENERGY CONSUMPTION BY REGION

While energy consumption is increasing in regions such as China and India, per-capita energy consumption continues to be highest in the developed world.

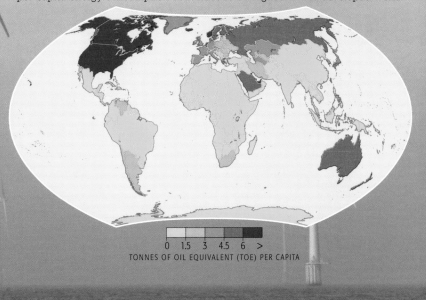

TONNES OF OIL EQUIVALENT (TOE) PER CAPITA

0   1.5   3   4.5   6   >

On the road again

Society currently relies almost exclusively on petroleum-based fuels, such as gasoline, for transport. This fuel use results in emissions of about 6 Gt CO_2 eq per year, and it is responsible for 13% of worldwide greenhouse gas emissions. Road vehicles produce the majority (about 75%) of this total. Over the past decade, emissions in the transport sector increased at an even faster rate than those in the energy sector. The greatest growth occurred in the area of freight transport (primarily by trucks for overland freight, and ships and airplanes for international transport). The rate of transport emissions is projected to increase even further over future decades, fueled by continued global economic growth and population increase.

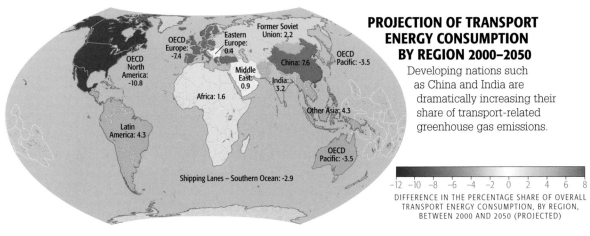

PROJECTION OF TRANSPORT ENERGY CONSUMPTION BY REGION 2000–2050

Developing nations such as China and India are dramatically increasing their share of transport-related greenhouse gas emissions.

-12 -10 -8 -6 -4 -2 0 2 4 6 8

DIFFERENCE IN THE PERCENTAGE SHARE OF OVERALL TRANSPORT ENERGY CONSUMPTION, BY REGION, BETWEEN 2000 AND 2050 (PROJECTED)

HISTORICAL AND PROJECTED TRANSPORT EMISSIONS BY MODE 1970–2050

While land-based modes of transport are likely to continue to dominate transport-related greenhouse gas emissions in the decades ahead, air travel is projected to make an increasingly large contribution.

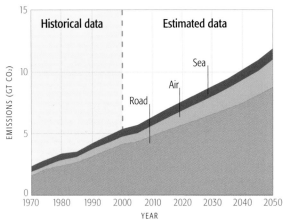

PROJECTION OF TRANSPORT ENERGY CONSUMPTION BY MODE 2000–2050

If we look at the projected trends more closely, we see that much of the future increase in transport-related energy consumption is likely to come from a combination of personal ("light duty") vehicles and freight trucks.

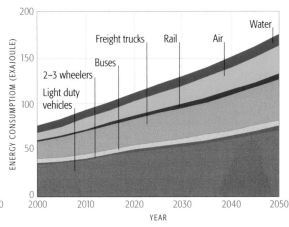

WARNING
Only to be used by
flex-fuel vehicles

Harvest
BioEthanol

BioEthanol
E85

This bioethanol pump at a supermarket in Norwich, UK, serves up biofuels in lieu of fossil fuels. Replacing conventional gasoline with biofuels could reduce transport-related greenhouse gas emissions.

While most people in the developing world still do not possess a personal vehicle, and many have no access at all to motorized transportation, this situation is changing dramatically. In the absence of a radical shift from current practices, transport-related carbon emissions are predicted to nearly double in the next few decades.

The potential of the so-called "Peak Oil" phenomenon (see p.160), to slow the future rate of growth has been greatly overstated. Even if conventional oil fields were to be depleted in coming decades, or drilling were to become prohibitively expensive, sources such as oil shales and tar sands could provide many additional decades of petroleum reserves. As coal liquification technology advances, coal could potentially satisfy the transport sector's rising fuel demands.

Emerging fuel-cell technology has often been cited as a potential solution. However, the use of fuel cells alone for passenger vehicles could simply shift the energy burden from the transport sector to the energy sector, with no net decrease in emissions. This is because fuel cells, which convert fuel energy into electricity in a manner similar to a battery, have to be recharged, and the energy to do that has to come from some place.

The quest for fuel efficiency

Meeting rising global transport sector energy demands while slowing the rate of fossil-fuel emissions will require a combination of greater fuel efficiency and increased use of carbon-free or carbon-neutral technologies. In the short term, this can be accomplished with more fuel-efficient vehicles, such as gasoline/electric hybrid cars and clean diesel vehicles, and through increased use of certain types of biofuels as gasoline additives or substitutes. Two decades from now, biofuels could satisfy 5–10% of the total transport energy demand. Technological innovations and improved air traffic management could result in better fuel efficiency in the aviation sector. Increased reliance on trains, buses, and other public transport, car-pooling, and non-motorized transportation (such as cycling and walking) could also help to curb emissions.

Electric Vehicle

Refueling station for electric vehicles
In an environmentally friendly future, we could see stations like these replacing gas pumps as electric vehicles replace today's inefficient gas-guzzlers.

"No regrets"

In many cases, measures to reduce fossil-fuel consumption constitute "no regrets" strategies. For example, using our automobiles less has the desirable added benefit of reducing traffic congestion and improving air quality. And decreased gasoline consumption results in the national security bonus of reducing our reliance on volatile regimes. Furthermore, improved vehicle efficiency measures lead to savings in fuel expenditures—savings that can be invested in other areas of the economy.

Long-term plans

Additional emerging technologies may allow for further reductions in transport-related greenhouse gas emissions in the coming decades. New types of biofuels, electric and hybrid vehicles with more powerful and longer-running batteries, and more efficient aircraft can all substantially contribute to long-term emission-reduction goals.

eling Station

Who can help?

Public policy measures can aid the mitigation of transport-related emissions in a number of ways. This is particularly true for nations or communities that are still in the process of establishing transportation systems. For example, well-thought-out urban planning and land-use regulations can provide better access to public transport, and make it more appealing as a commuting option. Thoughtful urban planning can also reduce commuting distances. Governments can establish, enforce, and, where necessary, raise mandatory fuel economy standards.

encourage the use of efficient vehicles. Some of the obstacles to the success of these measures are the persistent preference of many consumers for vehicles with poor fuel efficiency, targeted corporate advertising campaigns that reinforce these preferences, and lobbying efforts by car companies to keep fuel efficiency standards low. Any substantive reduction in the transport sector's fuel consumption will require not only proactive government policies, but also greater corporate accountability and personal acceptance of responsibility by

Building green

The commercial and residential buildings sector is a large emitter of greenhouse gases, accounting for nearly 4 Gt CO_2 eq in 2004. Actually, this sector is also a significant consumer of energy, so if we include this energy use, the buildings sector becomes one the largest emitters of carbon dioxide, emitting approximately 9 Gt CO_2 eq per year in 2004. Including the energy-consumption-related emissions is important as we consider the climate impact of approaches taken to promote energy efficiency in the buildings sector. Happily, many of the approaches taken to reduce emissions in this sector are technologically mature and provide benefits in addition to reducing emissions.

There are two basic ways the buildings sector can reduce its carbon footprint:

- Reducing energy consumption in construction and building operation
- Switching to low-carbon or carbon-free energy sources

Here we focus on the first of these strategies; alternative energy sources are addressed in "Keeping the power turned on" (see p.160).

Reducing energy consumption

The green building movement encourages efficiency in the design, construction, operation, and demolition of buildings, with the goal of enhanced human health and reduced impact on the environment. In the United States, this movement is exemplified by Leadership in Energy and Environmental Design (LEED). Certification points are awarded by LEED for sustainability and efficiency, as well as for the optimization of energy performance and the use and re-use of recyclable materials. A successful LEED building takes into account such important concerns as the availability of alternative transportation to the building (e.g., buses, trains), habitat preservation, and indoor environmental quality.

Reducing energy use in new buildings means reducing the heating and cooling loads. This can be accomplished through passive solar design (taking best advantage of available solar energy) and better insulation. High-efficiency lighting, appliances, and heating and cooling systems, and high-reflectivity building materials, and multiple glazing in windows can also markedly reduce a building's emissions.

Green renovation

Energy savings in new construction can exceed 75%, which bodes well for the future. However, buildings have long lifetimes, so most buildings in existence today will still be in use in 2030. This means that close attention has to be paid to the renovation of existing buildings, because that is where most buildings-related emissions reductions will be made.

Although these so-called "green" renovation techniques can involve considerable costs up-front, there are economic savings in the long term associated with energy-use reduction, and a number of co-benefits as well, including improved indoor air quality. Green building can also create jobs and new business opportunities, which in turn enhances economic competitiveness and energy security.

To enhance the lure of these long-term benefits, some measure of government intervention may be necessary. Appliance efficiency standards, new building codes, mandatory labeling and certification, energy efficiency quotas, and tax benefits for green construction are all options. Most of these mechanisms have a high cost-effectiveness, and in many cases benefits can be realized without costs.

Stemming the rising greenhouse gas emissions from the buildings sector will require a strong political commitment to green construction. This may take the form of governmental monitoring and the enforcement of codes and regulations. In the end, though, if the green building movement is successful, the interior space where we spend much of our time will be healthier and more comfortable, and have a minimal impact on the global environment.

Industrial CO₂ pollution

The image of a factory belching out smoke is engrained in our minds as the epitome of environmental pollution. Although many industries have taken significant steps to reduce pollution, thanks to its intense use of energy, the industry sector is still a major source of carbon dioxide and other greenhouse gases. And that source is growing: from 6 Gt CO_2 eq in 1971 to nearly 10 Gt CO_2 eq in 2004. Of these 10 Gt, approximately 5 Gt are from heat and power production and 5 Gt are from industrial processes, such as cement production and chemical processing. According to the A1B "middle of the road" fossil-fuel emissions scenario (see p.86), there will be little change in CO_2 equivalent emissions from the industrial sector by 2030.

CARBON CAPTURE AND STORAGE (CCS)

Carbon dioxide can be captured before it is released into the atmosphere and transferred underground via pipeline. Possible repositories include coal and salt beds, depleted oil and gas reservoirs, and saline aquifers.

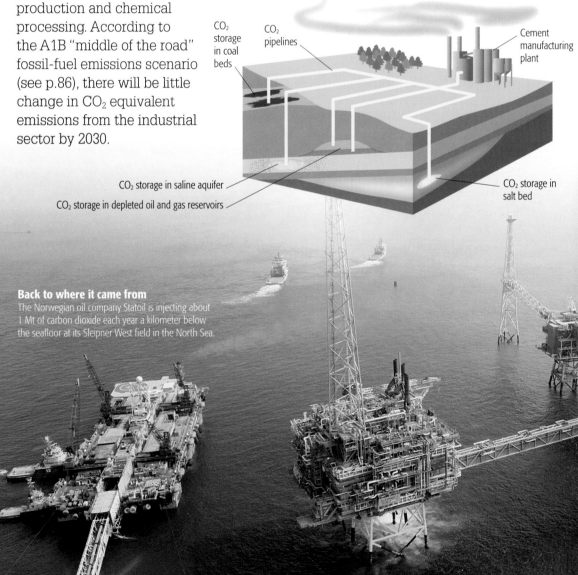

CO_2 storage in coal beds

CO_2 pipelines

Cement manufacturing plant

CO_2 storage in saline aquifer

CO_2 storage in depleted oil and gas reservoirs

CO_2 storage in salt bed

Back to where it came from
The Norwegian oil company Statoil is injecting about 1 Mt of carbon dioxide each year a kilometer below the seafloor at its Sleipner West field in the North Sea.

Industrial CO_2 emission is increasingly becoming an issue for developing nations. In 1971 only 18% of industry-related CO_2 emissions came from developing countries, but by 2004 their share had risen to 53%. This shift reflects both the growth of industry in developing countries and the movement toward improved energy efficiency in developed countries.

Industrial mitigation

There are numerous opportunities for mitigation of greenhouse gas emissions in industry, in part because many factories are using old and inefficient processes. Retrofitting of these factories—replacing electric motors and boilers, using recycled materials for fuel, and fixing leaks in furnaces and air and steam lines—could go a long way toward limiting carbon emissions in the future. Carbon capture and storage is another promising strategy to help realize industrial reductions.

REDUCING INDUSTRIAL CO₂ EMISSIONS THROUGH TAXATION & CARBON SEQUESTRATION

Tax: $20–50 per metric ton **Emission reduction:** 0.43–1.5 Gt CO_2/yr

CCS Cost: $20–30 per metric ton **Emission reduction:** 0.07–0.18 Gt CO_2/yr

STEEL
Total amount of projected CO_2 emissions by 2030: **1.2 billion metric tons (Gt)**

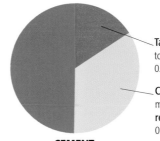

Tax: up to $50 per metric ton **Emission reduction:** 0.72–2.1 Gt CO_2/yr

CCS Cost: $50–250 per metric ton **Emission reduction:** 0.25–4.5 Gt CO_2/yr

CEMENT
Total amount of projected annual CO_2 emissions by 2030: **6.5 billion metric tons (Gt)**

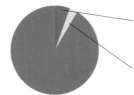

Tax: up to $20 per metric ton **Emission reduction:** 0.15–0.30 Gt CO_2/yr

CCS Cost: $17–31 per metric ton **Emission reduction:** 0.08–0.15 Gt CO_2/yr

PETROLEUM REFINING
Total amount of projected annual CO_2 emissions by 2030: **4.7 billion metric tons (Gt)**

Tax: up to $20 per metric ton **Emission reduction:** 0.05–0.42 Gt CO_2/yr

Carbon capture does not apply to this industry

PULP & PAPER
Total amount of projected annual CO_2 emissions by 2030: **1.3 billion metric tons (Gt)**

PROJECTED ANNUAL CO₂ EMISSIONS BY 2030
These values include emissions resulting from energy supplied to industry from the energy sector.

EFFECT OF CARBON TAX ON CO₂ EMISSIONS
If government agencies were to impose a tax of the amount listed (US $ per metric ton of emissions) on an industry, then it is estimated that the taxed industry would potentially reduce its annual emissions by the amount shown in red.

COST OF CARBON SEQUESTRATION
Some industries also have the potential to remove and store CO_2 from industrial emissions before it is released to the atmosphere. Different industries can sequester different amounts per year. If CCS procedures cost the amounts listed above (per metric ton of emissions), then it is estimated that each industry would potentially reduce its emissions by the amounts shown in yellow.

Greener acres

Given that nearly half of Earth's land surface is used for farming (crops and grazing), it should come as no surprise that the agriculture sector is a significant contributor to global greenhouse gas emissions. Farming and agriculture are responsible for annual emissions of about 6 Gt CO_2 eq. This is about 13% of worldwide greenhouse gas emissions, and roughly equal to the contribution from transport. Agricultural emissions have increased by 17% over the past 15 years. Much of that increase has come from the developing world, which is now responsible for about 75% of worldwide agricultural emissions. Interestingly, net CO_2 emissions from agriculture are negligible; plants do produce CO_2, but they consume it at about the same rate. The main agricultural emission is methane (CH_4), which is produced by microbes that thrive in environments such as rice paddies and the stomachs of ruminants like cattle, oxen, and sheep. Nitrous oxide (N_2O) from manure and other fertilizers is another agriculture-generated greenhouse gas.

Agricultural mitigation

Carbon is absorbed by and stored in living plants on Earth. Farming and grazing lands thus represent large potential carbon stores (places where carbon is sequestered away from the atmosphere). The land's ability to store carbon has decreased significantly over time, due to heavy-handed agricultural practices. Consequently, the greatest potential for agricultural mitigation lies not in the reduction of emissions themselves, but in the improved management of agricultural lands, which will restore their ability to sequester CO_2. Positive management practices include reducing soil tillage and restoring carbon-absorbing organic soils. More Earth-friendly practices that restore degraded land, such as converting aging crop land to grassland, can similarly aid in mitigation.

Of course, it is also a good idea to decrease agricultural emissions. Improved management practices can play a key role in this endeavor. For example, more efficient fertilizer delivery methods can minimize nitrous oxide emissions. Rice paddies can likewise be better managed to reduce methane production. And using alternative feeds will result in less methane production by ruminants. For example, recent experiments show that adding certain types of food-industry byproducts, such as cooking fat, to cattle feed reduced their methane production.

GLOBAL MITIGATION POTENTIAL OF VARIOUS AGRICULTURAL MANAGEMENT PRACTICES BY 2030

Better management of crop lands, grazing lands, and soils can decrease net greenhouse emissions by allowing land to more effectively sequester atmospheric carbon dioxide. Better farming practices can also reduce methane production, another important contributor to greenhouse gas emissions. The negative values on the graph indicate that rather than mitigating, the action in question is adding to current emissions.

KEY
Nitrous oxide/N_2O
Methane/CH_4
Carbon dioxide/CO_2

MEGATONS CO_2 EQUIVALENT PER YEAR

1800
1600
1400
1200
1000
800
600
400
200
0
−200

CROP LAND MANAGEMENT
GRAZING LAND MANAGEMENT
RESTORE CULTIVATED ORGANIC SOILS
RESTORE DEGRADED LANDS
RICE MANAGEMENT
LIVESTOCK
BIOFUEL
WATER MANAGEMENT
SET-ASIDE, LAND USE CHANGE & AGROFORESTRY
MANURE MANAGEMENT

Gathering rice husks
If farmers, like these in southern India, take good care of their farm lands, they can dramatically increase the land's ability to sequester CO_2 from the atmosphere.

Biofuel options

There is also potential for mitigation in agricultural production of biofuels. Burning biofuels does not lead to any net increase in greenhouse gas concentrations because the carbon released was just recently removed from the atmosphere and has only been stored in plants for a few months. For every gram of CO_2 released when a biofuel is burned, a gram was removed from the atmosphere by photosynthesis just a short time ago. This balance is why biofuels are considered "carbon neutral." Crops and agricultural residues can be used either as crude biofuel energy sources (e.g., burned for heat), or they can be chemically altered to yield more efficient biofuels such as ethanol or biodiesel.

Corn, one of the most widely grown cereal crops, can be readily converted into ethanol. However, there are at least two problems that limit prospects for the widespread use of corn-based ethanol as a fuel. First, there are troubling ethical considerations associated with the prospect of trading food for energy in this way, when starvation and malnutrition still afflict large numbers of people, especially in developing countries. Secondly, and perhaps more practically, the processes used to convert corn to ethanol are not very efficient; they require a considerable input of energy, limiting net gains. Researchers are currently seeking alternative pathways for ethanol production.

So-called "cellulosic ethanol," which can be derived from agricultural products such as switchgrass, could yield larger net gains in the future. (Switchgrass is a tall grass that grows naturally on the prairies of North America. Research has demonstrated that it has the potential to yield biofuel more efficiently than corn. In comparison with corn, switchgrass is more hardy with respect to soil and climate conditions, is perennial, and requires far less fertilizer and herbicide.)

Uncertainties

Most (approximately 70%) of the potential for mitigation in the agricultural sector lies in the developing world, specifically in China, India, and much of South America. It is challenging to determine precisely what the mitigation potentials of various options are. Key uncertainties include the willingness of governments to promote and support mitigation practices, and the willingness of individual farmers to adopt preferable management practices. Also uncertain is the continued effectiveness of various mitigation strategies in the face of escalating climate change, growing populations, and evolving technology.

Corn growing in Minnesota
Corn, which can be made into ethanol, is one possible source of biofuels. Other alternative biofuels, however, may yield greater mitigation benefits.

ESTIMATED MITIGATION POTENTIAL IN THE AGRICULTURAL SECTOR BY 2030

Note that South America, China, and India have the greatest mitigation potential.

0 200 400 600 800 1000

ESTIMATED MITIGATION POTENTIAL
(MT CO$_2$ EQ/YR) IN THE AGRICULTURAL
SECTOR FOR EACH REGION BY 2030

Forests
Source or sink for atmospheric CO_2?

Long before humans were burning fossil fuels, we were contributing to the buildup of carbon dioxide in the atmosphere. The practice of deforestation has accompanied human settlement and agriculture across the globe. The combustion of timber for energy and the gradual decay of lumber used in construction both release CO_2 into the atmosphere. In 2004, the forestry sector emitted roughly 17% of the total greenhouse gases released to the atmosphere.

Carbon uptake

During the pre-industrial era, forest-clearing and wood burning were common practices in Europe and the United States. Recently, however, previously cleared agricultural lands have returned to forests in these regions. As a result, forest-related CO_2 emissions have declined over the last several decades (see graphs opposite) and reforested lands have now become carbon sinks.

Now the developing world is repeating history, aggressively cutting down and burning trees (see map opposite). Deforestation in tropical South and Southeast Asia, Africa, and South America has recently accelerated. Between 2000 and 2005, an area roughly the size of Ireland was lost each year to deforestation.

This new carbon influx from tropical deforestation has been partially offset by the reforestation uptake in Europe, America, and elsewhere. Current emissions from deforestation amount to nearly 6 Gt CO_2 eq per year. If we include emissions from decomposition of logging debris, peat fires, and peat decay, forestry emissions add up to over 8 Gt CO_2 eq per year. On the flip side, it is estimated that approximately 3.3 Gt CO_2 eq per year are taken up through reforestation.

Obviously, the most expeditious way to reduce CO_2 emissions from the forestry sector is to prevent deforestation. But the developed world must keep in mind its own history.

Reforestation potential

What about mitigating via reforestation efforts? The efficacy of reforestation is contingent on favorable climate conditions. Deforested soils tend to dry out, and they are often low in nutrients because most of the ecosystem nutrients were stored in the harvested trees. For this and other reasons, reforestation of tropical rainforest has often been unsuccessful, and many experts believe that reforestation may not be a significant carbon sink in the second half of this century.

Timber!
This Indonesian tropical rainforest is being clear cut so that a palm-oil plantation can be established.

RATE OF CHANGE IN FORESTED AREA

This map shows the rate of change in forested area between 2000 and 2005. Note that the highest rates of deforestation (in red) are largely in the tropics.

< -0.5   0.5 >

NET CHANGES IN FORESTED AREA BETWEEN 2000
AND 2005 (PERCENTAGE CHANGE PER YEAR)

HISTORICAL TRENDS IN FOREST CARBON EMISSIONS AND UPTAKE

These graphs show historical trends in forest carbon emissions (red) and uptake (green), for the period between 1855 and 2000, in Mt CO$_2$ eq. The US and Europe have become net carbon sinks after a long history of deforestation.

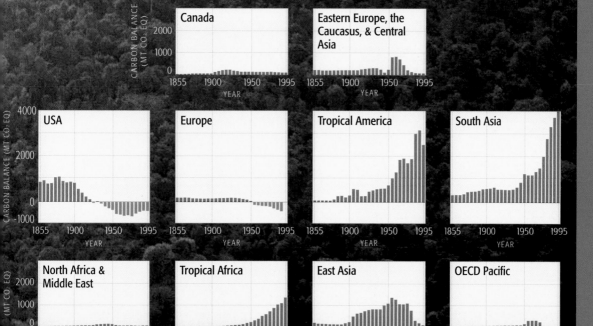

Waste

Life pollutes: humans pollute, cats pollute, ferns pollute, and bacteria pollute. Organisms metabolize food to gain the energy they need to grow, reproduce, and move. Metabolism creates waste products; these are often toxic and need to be eliminated. Although some level of pollution is unavoidable, humans pollute for reasons other than just simple metabolism.

Landfill problems

Although waste disposal is a significant problem facing an ever-expanding and consumptive world population, greenhouse gas emissions from the waste management sector account for only about 3% of total global emissions (1.3 Gt CO_2 eq per year). The largest share of waste-related emissions comes from landfills, primarily in the form of methane. The bacteria that decompose waste in the oxygen-depleted interior of landfills produce methane. Well-aerated landfills produce carbon dioxide instead of methane, which molecule-for-molecule is a weaker greenhouse gas.

Researchers digging in landfills have exhumed decades-old hot dogs that showed no signs of decomposition. In the context of the carbon cycle, those hot dogs, and much of the other waste accumulating in typical, inefficient landfills, are carbon "sinks." To some extent carbon sinks offset the release of carbon dioxide from fossil-fuel burning. For example, a corn plant may take up a carbon dioxide molecule recently emitted from a coal-fired power plant. The corn is then fed to a pig, and the pig is slaughtered to make hot dogs. A hot dog is discarded by a child at the ball game, and ends up in a landfill where it may sit for centuries without decomposing and releasing its carbon back into the atmosphere. Taken together, all the hot dogs and the rest of the 0.9 Gt CO_2 eq of waste generated globally each year "store" about 0.2 Gt CO_2 eq. This number is small, even compared to the relatively low total waste sector emissions, so we certainly don't want to produce more waste just to sequester a small amount of carbon.

Is that a kitchen sink or a carbon sink?
Landfills like this one in Tokyo Bay, Japan, both
accumulate carbon wastes and emit greenhouse gases.

Energy from waste

Rather than attempting to reduce already
low waste sector emissions further, some
communities are exploring waste recycling
alternatives, including burning landfill
garbage to use as a renewable energy
resource (as a fossil-fuel substitute). Society
is already obliged to collect waste and
transport it to a central repository; why not
burn it for inexpensive energy rather than
just allowing it to accumulate and release
methane? State-of-the-art technology
is able to prevent much of the potential
pollution from incineration. The methane
seeping out of landfills can also be captured
and used for energy production, further
reducing the overall impact of landfills on
the environment. This large-scale garbage
"recycling" may turn out to be a win–win
situation for society as we struggle to find
ways to mitigate emissions and efficiently
manage waste.

WASTE SECTOR EMISSIONS PROJECTION THROUGH 2050, AND THE EFFECT OF VARIOUS MITIGATION STRATEGIES

The emission of greenhouse gases from landfills
and other waste-disposal activities will increase
dramatically over the next 40–50 years, but
increased incineration of waste (to use as a
substitute for fossil-fuel energy), and expanded
efforts to capture landfill methane can slow
the increase.

EMISSIONS (GT CO₂ EQ)

3.0
2.5
2.0
1.5
1.0
0.5

Without mitigation

Increased
incineration

Increased
recycling

Increased methane capture

Geoengineering
Having our cake and eating it too

Geoengineering is an alternative approach to mitigation that involves using technology to counteract climate change impacts either at the source level (doing something about growing greenhouse gas levels) or at the impact level (offsetting climate change itself). These approaches involve planetary-scale environmental engineering the likes of which society has never before witnessed.

Carbon sinks

One source-level geoengineering proposal, referred to as "iron fertilization," involves adding iron to the upper ocean. Iron is a nutrient that is of limited availability in the upper ocean. This scarcity of iron places limits on the activity of marine plants that live near the ocean surface. Some scientists think that iron fertilization can increase the rate at which plants in the upper ocean take up carbon dioxide, thus boosting the efficiency of the deep-ocean carbon sink (see p.94), and offsetting the buildup of carbon dioxide in the atmosphere. However, the limited experiments that have been done suggest that the main effect of iron fertilization would probably simply be faster cycling of carbon between the atmosphere and the upper ocean, with little or no burial of carbon in the deep ocean. In addition, there could be possible negative side effects if humans interfere further with the complex and delicate ecology of the marine

biosphere. Other geoengineering approaches include attempts to increase the efficiency of terrestrial carbon sinks by planting more trees and "greening" regions that are currently deserts. Many consider this approach far more environmentally friendly than other proposed schemes, but it is unclear if it could be accomplished on the massive, planetary scale required to significantly offset human carbon emissions.

Carbon capture

Closely related to regional greening plans are so-called carbon capture and sequestration (CCS) approaches. In CCS approaches, carbon is extracted from fossil fuels as they are burned, preventing its escape and buildup in the atmosphere. The captured carbon is then buried and trapped well beneath Earth's surface or injected into the deep ocean, where it is likely to reside for many centuries. One potentially effective CCS scheme would involve scrubbing carbon dioxide from smokestacks, and reacting it with igneous rocks to form limestone. This mimics the way that nature itself removes carbon dioxide from the atmosphere over geological timescales (see p.94). Recently, Klaus Lachner of Columbia University has argued for a related alternative, in which massive arrays of artificial "trees" take carbon directly out of the air and precipitate it in a form that can be sequestered.

Saltwater in the sky
This artist's conception shows a proposed device for spraying large quantities of sea water into the atmosphere to help boost the sun-reflecting power of marine stratocumulus clouds.

Solar shields and aerosols

One of the most frequently proposed impact-level geoengineering approaches involves deliberately decreasing the amount of sunlight reaching Earth's surface to such a degree that the reduction in incoming radiation offsets any greenhouse warming. One such method involves deploying vast "solar shields" in space that reflect sunlight away from Earth. Shooting sulphate aerosols into the stratosphere to mimic the cooling impact of volcanic eruptions (see p.18) is a less costly, but potentially more dangerous alternative. This method could exacerbate the problem of ozone depletion (see p.30) by tampering with the chemical composition of the stratosphere.

While calculations suggest that either of these impact-level methods could indeed offset greenhouse warming of the atmosphere, they each come with problems of their own. First, they do nothing to avert the problem of ocean acidification associated with increasing atmospheric carbon dioxide levels (see p.114). Furthemore, climate models indicate that reducing the incoming solar radiation, while potentially offseting the warming of the globe, would not necessarily counteract the regional impacts of greenhouse warming. Some regions might warm at even greater rates, and patterns of rainfall and drought could be dramatically altered. Not to mention that if, for some reason, these activities were ultimately halted, the

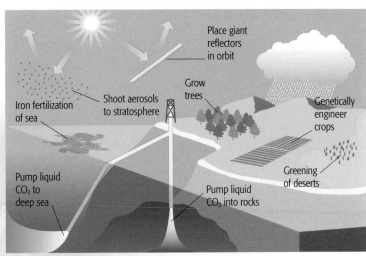

Various geoengineering schemes, such as the ones illustrated here, have been proposed to offset the impacts of fossil-fuel burning on climate.

full impact of warming that had remained hidden for decades would suddenly be unmasked, leading to dramatic, rapid global climate change.

Schemes of last resort

Each of the proposed geoengineering schemes has possible shortcomings and poses a potential danger. Some advocates maintain that if we are backed into a corner and faced with the prospect of irreversible and dangerous climate change, we may need to resort to these schemes at least as partial solutions. Others note that it would be wise not to tamper with a system such as the climate, the workings of which we still do not entirely understand. Either way, the debate over whether geoengineering is likely to be an effective and prudent solution to climate change is bound to continue—as scientists will continue to come up with new proposals for using technology to address climate change problems.

But what can I do about it?

If all this talk about energy sectors and governmental buy-in leaves you feeling helpless in the face of global warming, don't let it! We can all make lifestyle choices that will directly aid in the mitigation of greenhouse gas emissions. In many cases, these are "no regrets" changes that have positive side benefits, such as improving our health and quality of life, conserving natural resources, and facilitating greater environmental sustainability.

Lifestyle choices

First and foremost, we can be more efficient in our use of energy. We can save energy daily by making home improvements that decrease the energy we use to heat and cool our houses and apartments. More efficient practices include better insulation, passive solar heating, and the substitution of fans and open windows for air conditioning when practical. We can replace inefficient incandescent light bulbs with more efficient, compact fluorescent bulbs. A tremendous amount of energy is used in industrial manufacturing processes (see p.168), so there is also significant mitigation opportunity in simply being better about recycling.

There are other lifestyle changes we can make easily and immediately, ones that don't require that we remodel or even buy new bulbs or appliances. For example, in warm and dry weather, clotheslines make an excellent substitute for dryers. Appliances that are not in use can be unplugged, reducing electricity leakage.

We can make serious contributions to emission reduction efforts with our transportation choices. Many of us could commute to work by bicycle or on foot. For those of us who have difficulty finding time to maintain fitness regimes, this option allows for the best sort of multitasking—we exercise while we reduce our carbon footprints.

Other alternatives to driving long distances alone include public transportation and carpools. Fuel-efficient vehicles are another exciting new option. Given the high cost of gasoline in recent years, this option not only benefits the environment, but our pocketbooks as well.

Education and incentives

In addition to what individuals can do, employers, governments, and non-governmental organizations can play an important role. Community-focused organizations can provide relevant guidance and education to individuals. Some governments already provide tax benefits and incentives for citizens who build green, add solar panels to their roof, or buy hybrid vehicles. Public outreach efforts can also include educational programs that teach energy conservation practices, and campaigns aimed at encouraging individuals to make environmentally conscious decisions. If you want to know how well you are currently doing in terms of your own personal contribution to global greenhouse gas emissions, turn to p.182.

Decrease the amount of energy used in your home by installing simple solar panels.

Appliances that are
not in use can be
unplugged, reducing
electricity leakage.

Clotheslines make
an excellent substitute for dryers.

ecodrive
ELECTRIC VEHICLES

Drive alone less or drive
a fuel-efficient or
electric vehicle.

Carpool
Only

Remember
to recycle.

Replace incandescent bulbs with
fluorescent bulbs.

Commute to work
by bicycle or on foot.

What's your carbon footprint?

Before you go on a diet, you might weigh yourself to establish a starting point and also, perhaps, to further motivate yourself to cut back on calories. In a similar way, you can calculate your "carbon footprint"—your personal contribution to the problem of global warming—as a first step in reducing the emissions your lifestyle generates.

There are numerous carbon footprint calculators on the Internet; we provide a few Web site addresses at the end of this discussion for your reference. All of these calculators help you to evaluate your lifestyle and determine your personal contribution or "footprint." Carbon footprints are usually measured in metric tons of CO_2 eq per year.

In order to assess your footprint, you will be asked to answer a series of questions about your lifestyle.

Footprint questions

- Where do you live? There are regional differences in the energy sources used to generate electrical power and these affect emissions.

- How many people live in your home? Your carbon footprint is smaller if the energy you use is shared.

- What type of vehicle and how many miles per year do you drive? Your car's gas mileage affects emissions.

- How often do you fly, and are your trips short or long? Airline travel is a big emissions contributor. Short trips use more energy per mile because takeoffs are particularly fuel-intensive.

- Do you heat your home with natural gas, heating oil, or propane? How much is your typical monthly heating bill? What is your typical monthly electric bill?

- Do you eat red meat, just chicken and fish, or are you a vegetarian? The various activities that put these foods on your table have differing carbon emission profiles.

- Do you eat mostly local foods and buy mostly local products, or do you prefer imported goods? Long-haul freight consumes fossil fuel aggressively.

- What types of recreation do you prefer? A bicycling trip that begins at your own front door contrasts quite markedly with snowmobiling or motorboating at distant recreation sites.

Now take off your shoes and compare your print to two famous footprints!

SASQUATCH

Carbon footprint:

30 METRIC TONS CO_2 EQ PER YEAR

At home
Lives in a poorly insulated, air-conditioned apartment in the American Midwest with electric heat; loves to soak in a hot bathtub several times a week; does not recycle.

On the go
Drives an SUV 24,000 km per year; takes 5–10 business trips by plane each year, both short haul and long haul; and takes one exotic personal vacation each year by plane.

CINDERELLA

Carbon footprint:

4 METRIC TONS CO_2 EQ PER YEAR

At home
Lives with two roommates in France in an energy-efficient, well-insulated apartment with electric heat; prefers taking showers; recycles a significant fraction of household waste.

On the go
Drives a hybrid car 16,000 km per year; journeys by train 1600 km per year; uses videoconferencing rather than traveling for business; and takes one exotic personal vacation each year by plane.

Does your footprint looks more like Cinderella's dainty glass slipper or Sasquatch's big paw? Any of the following Web sites will help you get closer to the answer and suggest ways in which you can become more carbon stingy and environmentally friendly.

ON-LINE CARBON FOOTPRINT CALCULATORS

http://www.epa.gov/climatechange/emissions/ind_calculator.html
http://atmospheres.gsfc.nasa.gov/iglo/
http://www.carbonfootprint.com/
http://www.climatecrisis.net/takeaction/carboncalculator/
http://www.bp.com/extendedsectiongenericarticle.do?categoryId=9015627&contentId=7029058

Global problems require international cooperation

The atmosphere does not recognize national boundaries. When pollutants such as greenhouse gases or industrial aerosol particulates are emitted, they travel great distances, crossing continents and oceans. No single nation can solve the problems created by atmospheric pollutants. Some argue that, because of its pervasiveness, global warming is too daunting a challenge, that humans simply cannot solve a problem so huge in scale—especially one that requires cooperation between so many disparate parties. History, however, suggests otherwise, and even provides precedent for cooperation between nations in solving environmental problems.

Acid rain

By the 1970s, large parts of the northeastern United States and eastern Canada were plagued by the problem of "acid rain." As lakes, rivers, and ponds became acidified, fish populations and other aquatic life died off in startling numbers. Acidic rainfall began to kill trees as well. Research implicated the sulphate and nitrate aerosol particulates produced by factories located

in the American Midwest. Aerosols were being carried downwind, where they dissolved in rainfall to form sulfuric and nitric acid. This so-called "acid rain" ultimately ended up in streams, rivers, ponds, and lakes. Europe also experienced similar acid rain problems. In some cases, the acid rain was destroying historic monuments and structures, as well as natural habitats.

In response to these and other related problems, the United States passsed a series of laws, culminating with the "Clean Air Act" of 1990. This act included specific provisions for dealing with the acid rain problem. The Clean Air Act of 1990 (and other clean air acts passed in the US and other nations) led to the widespread introduction of "scrubbers" into factories, which remove harmful particulates from industrial emissions before they enter the atmosphere.

The effects of acid rain
This conifer forest on Mount Mitchell, in North Carolina, includes many trees damaged or killed by acid rain.

Ozone depletion

Perhaps an even better analogy to the challenge posed by climate change is the problem of ozone depletion (see p.30). A breakdown in the stratospheric ozone layer was measured at the South Pole in the early 1980s. Theoretical and observational considerations pointed to an anthropogenic cause. Industrial products such as chlorofluorocarbons (CFCs), used at the time as refrigerants and propellants in aerosol cans, were eventually reaching the stratosphere. There, in the presence of solar radiation, they produced chemicals capable of destroying the ozone layer.

Since the ozone layer prevents most harmful ultraviolet radiation from reaching Earth's surface, the depletion of this protective layer was tied to an increase in skin cancers, and other damaging effects on plants and animals. In 1989, worldwide concern led to adoption of the Montreal Protocol, an international agreement banning the production of ozone-depleting chemicals. Former UN Secretary-General Kofi Annan has referred to the protocol as:

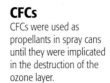

CFCs
CFCs were used as propellants in spray cans until they were implicated in the destruction of the ozone layer.

"Perhaps the single most successful international agreement to date."

Former UN Secretary General Kofi Annan

A global challenge

Arguably, the problem of climate change is more challenging to solve than that of ozone depletion. In that case, other commercial refrigerants and propellants were readily available as substitutes for ozone-depleting substances. By contrast, the emission of greenhouse gases results from the world's dependence on fossil fuels—its primary source of energy—and unfortunately a substitute for fossil fuels that can meet current (and future) world energy demands has yet to be found. Emission control and reduction requires a fundamental change in global energy policies.

The Kyoto Protocol

Recent efforts to achieve just that have begun:

- **1992:** The United Nations Framework Convention on Climate Change (UNFCCC), an international treaty, was formulated at the "Earth Summit" held in Rio De Janeiro, Brazil.

- **1997:** Five years later, an update to the UNFCCC—the " Kyoto Protocol"—was agreed upon at a summit in Japan.

- **2005:** Only after another eight years did the treaty actually go into force. The stated objective of the Kyoto Protocol is to achieve "stabilization of greenhouse gas concentrations in the atmosphere at a level that would prevent dangerous anthropogenic interference with the climate system."

- **At present:** 172 parties have currently ratified the Kyoto Protocol, including all but two major industrial countries. All of the parties that have entered into the agreement are committed to reducing greenhouse gas emissions to mandated levels. The two industrialized nations that have not ratified the protocol are the United States and Australia. However, the Australian Prime Minister elected in November of 2007 has pledged to sign. A large number (137) of developing nations have also ratified the protocol, but they are not held to mandated reductions in light of the financial hardships doing so might impose upon their fragile economies. The protocol includes provisions to insure that the developed world assists developing nations in moving toward more environmentally friendly energy resources.

The Kyoto Protocol has been criticized by both sides in the climate change debate. Critics on one side argue that it doesn't go far enough, and that the emission cuts mandated in the protocol will not stabilize greenhouse gas concentrations below dangerous levels. Supporters point out, however, that it is just a first step, putting in place a framework that can be built upon in the future to achieve broader reductions. Critics on the other side argue that committing to Kyoto will destroy the global economy. Yet cost-benefit analyses suggest that the cost of inaction could

Logging concessions

A timber truck transports logs in the Miri interior, eastern Malaysian Borneo state of Sarawak. A group of developed nations and an American green group donated U.S. $160 million for a World Bank-led climate-change plan in December 2007, which encourages developing nations to conserve their tropical forests. The project was launched in Bali, Indonesia, amid negotiations to establish a new global climate change agreement for 2012 and beyond.

be far greater (see p.146). The current debate is arguably more about politics than objective scientific or economic considerations; as such, it is likely to continue into the forseeable future.

Post-Kyoto period

Many feel that an additional breakthrough is still needed to produce an effective global agreement to stabilize greenhouse gas emissions. An obvious timeframe for this would be 2012, when the commitment period for the Kyoto Protocol expires. In December of 2007, further negotiations took place at a conference in Bali, Indonesia. The purpose of this conference was to establish a new global greenhouse emissions agreement for the post-Kyoto period, lay groundwork for the required international negotiations, and begin to establish a timeline for finalization of the new agreement. The various delegations, including major players such as the US, the European Union, China, Japan, and Canada, agreed to a "roadmap" in Bali for future negotiations. However, certain delegations (e.g., the US, Canada, and Japan) opposed specific cutback targets championed by the EU and others. As a result, the roadmap is a compromise between the competing interests of different nations. The UK Prime Minister, Gordon Brown, called the agreement "a vital step forward for the whole world." However, he warned that the agreement was "just the first step...now begins the hardest work, as all nations work towards a deal in Copenhagen in 2009 to address the defining challenge of our time."

Climate change summit
A new roadmap for addressing human-caused climate change was reached at the UN Summit in Bali in December 2007.

united nations climate change conference
Nusa Dua - Bali, Indonesia, 3-14 December 2007

Can we achieve sustainable development?

A responsible society strives to meet its needs without compromising the ability of future generations to meet theirs. This defines sustainable development. Sustainability requires that we protect ecosystems from destruction and consume natural resources at a rate no greater than nature can provide. Achieving environmental sustainability is difficult in terms of water use and soil erosion, and seemingly impossible when we take into account fossil-fuel and mineral-resource consumption. Since cutting consumption is such a challenge, we should also look to renewable energy and recycling to achieve sustainability goals.

Developing nations

Issues of equity enter into the sustainability equation because current models of economic development depend on increased consumption and depletion of natural resources. The challenge is to find ways that developing countries can achieve a quality of life equal to that of the developed world without damaging the environment and depleting resources. Doing so requires that they make a substantial shift away from the highly consumptive and largely unsustainable path followed by the developed world. Fortunately, developing nations can now utilize previously unavailable technologies that may help them to meet their needs with reduced impact on the environment.

For example, China has slowed its fossil-fuel-use increases with a combination of activities. By shifting to renewable and less carbon-intensive energy sources, imposing economic reforms, and slowing population growth they have moved in a positive direction. India, Turkey, Mexico, South Africa, and Brazil are also working to decouple economic development from fossil-fuel dependency.

SUSTAINABLE DEVELOPMENT STRATEGIES

There are many strategies for mitigating against climate change. Most of these enhance sustainability but also involve trade-offs.

Mitigation option		
Improving energy efficiency	**Reforestation**	**Deforestation avoidance**
Compatibility with sustainable development — Cost effective; creates jobs; benefits human health and comfort; provides energy security	Slows soil erosion and water runoff	Sustains biodiversity and ecosystem function; creates potential for ecotourism
Trade-offs	Reduces land for agriculture	May result in loss of forest exploitation income and shift to wood substitutes that produce more emissions

In many cases, sustainable development strategies are clearly win-win, as is the case with alternative energy or carbon sequestration (see p.178). These technologies can enhance national security by reducing dependency on foreign oil, create new jobs, and stimulate economies (see table below). But in other instances, short-term economic benefits may conflict with environmental benefits. For example, the shift in developing countries from biomass (wood-fire) cooking to the use of cleaner and more efficient liquid propane (fossil-fuel) stoves enhances human health and quality of life by reducing indoor pollution, but increases dependency on fossil fuels. This, in turn, increases greenhouse gas emissions and exacerbates human-induced climate change.

Developing policy

The responsibility for implementing sustainable development policies lies with government, industry, and civil society:

- Just as human health has improved globally thanks to diverse local strategies, progress toward a common goal such as climate change mitigation can be made via disparate governmental policies.

- Sustainability and profitability are gradually being seen as compatible goals in industry, perhaps essentially so for large, multinational companies. Regulatory compliance is also a factor, but may not be as important as was once thought.

- Non-governmental organizations (NGOs), which encourage reform, provide policy research and advice, and champion environmental issues, are increasingly expressing the will of civil society. Academia also plays a role, especially in research, which enhances understanding of the scientific, economic, and political implications of climate change. The IPCC itself is a prime example of how a civil body can influence the world's response to global warming.

Fortunately, the goals of climate change mitigation and sustainable development are largely compatible. Together, these two strategies can help us to create a healthier and more durable society for the future.

Incineration of waste	Recycling	Switching from domestic fossil fuel to imported alternative energy	Switching from imported fossil fuels to domestic alternative energy
Energy is obtained from waste	Reduces need for raw materials; creates local jobs	Reduces local pollution; provides economic benefits for energy-exporters	Creates new local industries and employment; reduces emissions of pollutants; provides energy security
Air pollution prevention may be costly	May result in health concerns for those employed in waste recycling	Reduces energy security; worsens balance of trade for importers	Alternative energy sources can cause environmental damage and social disruption, e.g., hydroelectric dam construction

The ethics of climate change

The international media has paid considerable attention to the economic implications of global climate change. They have, by contrast, paid little attention to the equally important ethical considerations. The objective of the Kyoto Protocol itself—to stabilize "greenhouse gas concentrations...at a level that would prevent dangerous anthropogenic interference with the climate system"— begs several questions. Who, for example, determines what constitutes "dangerous"? Answering such questions requires us to take into account political, cultural, and philosophical principles that are fundamentally ethical in nature.

Winners and losers

One tricky ethical principle is "equity." Equity issues surrounding climate change include the fair distribution of risks, benefits, responsibilities, and costs to both developed and developing nations. Climate change will be associated with potentially dramatic redistributions of wealth and resources, impacting food production, fresh water availability, and environmental health. In the course of this shuffling, there will be winners and losers. Unfortunately, climate change won't play fair. In fact, climate change may play the role of a "reverse Robin Hood," taking resources from the poor, and giving them to the rich. Tropical regions—the developing world essentially—will likely suffer the most detrimental impacts. In the short term, the developed nations may even stand to benefit. Europe and North America agribusinesses, for example, may enjoy longer growing seasons (see p.130).

Further ethical complications arise from the fact that the individuals who gain from current fossil-fuel burning are not the same as the individuals who stand to lose when the climate changes. Is it possible to assign a meaningful cost to the devastating impacts of climate change on the poor and disadvantaged?

After the flood
A Bangladeshi woman collects water from a well submerged by flood water at Paikpara. Hundreds of people in the northern districts of the country sought shelter after their houses were lost in the severe flood of July 2007.

What is the value of the life of a starving child in Bangladesh as measured in cheap barrels of oil? Do we even dare to pose such questions?

And the developing world, by virtue of its relative poverty and lack of technological infrastructure, is far more vulnerable to the economic, environmental, and health threats posed by climate change. Ethical considerations would seem to demand that the developed world assist developing nations in adapting to climate change, both in mitigating impacts, and exploiting possible benefits.

The developed world has already benefited from a century of cheap fossil-fuel energy. Given this fact, it is surely unfair to tell developing nations, who are just now beginning to build their energy and transportation infrastructures, that they can't have their turn to enjoy cheap oil. This challenging ethical dilemma complicates discussions about the appropriate burden of mitigation, including the distribution of emissions rights both among nations and between generations.

Social discounting

When it comes to the generational transfer of the benefits and costs of fossil-fuel burning, "social discounting" places a greater value on benefits today at the expense of subsequent generations. This is based on the assumption that future generations will have access to new technology and will be better equipped to deal with environmental challenges. Social discounting involves making an ethical call. If we discount future potential impacts too strongly and assume that future generations will be able to solve all problems, the cost-benefit analysis will surely favor inaction. Is it fair to gamble like this, knowing that it is our grandchildren who will pay the price if our assumptions turn out to be wrong?

Geoengineering dilemmas

Ethical considerations are also raised by geoengineering approaches to mitigation (see p.178). Should nations that stand to benefit from certain types of interference be able to do so, even when other nations may be negatively impacted by their actions? Ethical issues complicate discussions of biofuel technology too. Should agricultural land currently used to feed people be reallocated for energy production at a time when starvation and malnourishment are omnipresent?

An essential step in tackling these problems is for all countries, including holdouts like the United States, to join the international effort to stem the buildup of greenhouse gases initiated by the Kyoto Protocol (see p.186).

The known unknowns and the unknown unknowns

There are at least two kinds of unknowns. There are the "known unknowns," which are the questions we already know to ask, but for which we don't yet have the answers. Then there are the "unknowns unknowns." These are the questions we don't even know to ask, the questions involving phenomena that currently lie beyond the horizons of our imagination.

A great deal of discussion in this book is devoted to the known unknowns. We have discussed the open scientific questions regarding how much warming is to be expected and precisely what the pattern of climate change will be. These uncertainties are linked to unknowns regarding the societal and environmental impacts of climate change (e.g., changes in water availability, food supply, and disease prevalence). We have examined the still unsolved mysteries of the great climate changes in Earth's past, and the changes in violent weather phenomena, such as hurricanes, that may lie in store for us in the future.

More known unknowns

The known unknowns also include the lack of certainty regarding the "tipping points" looming in our future. Scientists recognize that such tipping points probably exist, but they don't know exactly where they may lie:

- Just how rapidly will the major ice sheets melt, and how high will the sea level rise accordingly?

- Will the "conveyor belt" ocean circulation weaken? And if so, when?

- Will the ability of the oceans and plants to absorb the CO_2 we are adding to the atmosphere change in the future?

Also included in the known unknown category are answers to questions relating to the unpredictability of human behavior.

- What will future human-driven emissions patterns be?

- What will the economic implications of warming be?

- What steps will we take to mitigate against greenhouse gas buildup and climate change? How successful will mitigation efforts be?

- Will we implement any of the currently conceived geoengineering plans? Will new, risk-free plans be conceived of?

Unknown unknowns

And what about the unknown unknowns? There are some of these in the science itself:

- Will the response of the climate to increased greenhouse gas concentrations take an unpredicted course?

- What are the tipping points that have not been conceived of yet?

- Are there hidden reserves of carbon on our planet that could suddenly be released, leading to further warming?

In the case of adaptation and mitigation, the unknown unknowns may be the stuff of science fiction. Decades ago, who would have imagined modern-day technology such as cloning or hand-held "smart phones" as powerful as the supercomputers of previous decades? More to the point, who would have conceived of modern transportation options such as hybrid vehicles, or prospective energy technology such as "cellulosic ethanol" (see p.172)?

So what are we to make of all of this uncertainty?

Clearly, we must work to diminish the uncertainty where possible, particularly when it impacts on our ability to make appropriate policy decisions or choose an optimal strategy for mitigating climate change. Recent history has taught us that uncertainties are not adequate justification for avoiding action. We know enough today to understand how vital it is that we act now.

A scuba diver explores a dark underwater cave
What surprises are in store for us as we continue to probe the climate system's sensitivity to human insult?

The urgency of climate change
Why we must act now

Uncertainty abounds (see p.192) but it is a poor excuse for inaction. In fact, given the possibility of severe and irreversible harm to society and the environment, scientists generally advocate that we abide by a "precautionary principle" that puts the onus of proof on those advocating inaction.

No excuses

If the remaining uncertainty in the science is not a valid argument against taking immediate action to slow climate change, then what, if anything, is? As we have seen, some argue that action could harm the economy. Yet this argument does not appear to withstand scrutiny, since the economic harm of inaction looks to be greater (see p.146) in the long term. Others argue that climate change might be beneficial to humankind, but an impartial assessment strongly suggests otherwise (see p.104). Still others concede that climate change represents a potential threat, but that it is only one of many problems facing society, and that focusing on climate change issues might divert attention and resources from more pressing problems. The argument that we must choose between competing societal problems, however, is based on the flawed premises that society can only solve one problem at any given time, and that the problems facing society are independent of each other. In the case of climate change, we have already shown its potential to exacerbate other major global societal and environmental issues, including:

- Sustainability
- Regional conflict
- Biodiversity
- Extreme weather events
- Water availability
- Disease

Some proponents of inaction argue that we can engineer our way out of the problem with future technological "fixes." However, the potential pitfalls of high-tech fixes present risks as well. And many of these fixes may not be able to prevent or reverse the more serious consequences of climate change, such as the melting of the Greenland ice sheet. While there are some promising new technologies on the horizon, there are none currently available that will handily satisfy our global thirst for carbon-free or carbon-neutral energy in the decades to come. Furthermore, while carbon sequestration (CCS) may ultimately slow the buildup of carbon dioxide in the atmosphere, the feasibility of large-scale implementation of CCS has yet to be demonstrated (see p.178).

Climate change has been described as a problem with a huge "procrastination penalty." With each passing year of inaction, stabilizing Earth's climate becomes increasingly difficult.

Our children's world

If we choose not to act on this problem now, then in the very best-case scenario we must accept that our children and grandchildren will grow up in a world lacking some of the beauty and wonder of our world. They may come of age in a time where:

- Polar bears, golden toads, and numerous other creatures will be the stuff of myth

- There will be no Great Barrier Reef to explore

- Giraffes and elephants will no longer loom in the foreground of the majestic snows of Kilimanjaro

- Great coastal communities such as Amsterdam, Venice, and New Orleans will join the lost city of Pompeii

Of course, humankind might plausibly adapt to these sad changes.

In the worst case scenario, however, our grandchildren will grow up, as renowned climate scientist James Hansen has bluntly put it, on "a different planet"—a planet potentially resembling the dystopian world depicted in science fiction movies such as *Soylent Green* and *The Island.* Adaptation, in this case, is unlikely to be viable for many of the world's people and other living things.

A polar bear sits on a small iceberg
Will these amazing creatures be an early casualty of human-caused climate change?

195

Where does that leave us?

There is no "silver bullet" that will solve the problem of global climate change. But that does not mean we should throw up our hands in the face of this urgent problem. Any viable solution is going to require action from many governments and all strata of society; it will involve adapting to the changes that are inevitable, and mitigating the changes we can avert. It goes without saying that alternative energy sources must be aggressively developed and deployed, and that governments must incentivize and reward responsible behavior by individuals and corporations.

The future in our hands
Our planet has supported life for billions of years, but only over the past century has a species—humans—developed the ability to alter the planetary environment. Will we do good or harm with this newfound ability? The answer is in our hands.

Climate change is one of the greatest, if not **the greatest** challenge ever faced by human society.

But it is a challenge that we must confront, for the alternative is a future that is unpalatable, and potentially unlivable. While it is quite clear that inaction will have dire consequences, it is likewise certain that a concerted effort on the part of humanity to act in its own best interests has great potential to end in success.

Glossary

Acid rain

Acid rain refers to any form of precipitation (e.g., rain, snow, sleet) that is unusually acidic. Largely caused by industrial emissions of sulfur and nitrogen aerosols, which form sulfuric and nitric acid when combined with water droplets suspended in the atmosphere, acid rain has caused documented damage to trees and plants, fish and other aquatic animals, building facades, and monuments. In recent decades, the governments of the US, the UK, and Canada have passed legislation, such as the Clean Air Acts, to reduce the industrial emissions responsible for acid rain.

Aerosols

Aerosols are microscopic liquid droplets, dust, or particulate matter that are airborne in the atmosphere. An aerosol may remain suspended in the atmosphere for hours or years depending on the type of aerosol and its location in the atmosphere. Aerosols can be of either human or natural origin and they may reflect and/or absorb incoming and outgoing radiation. Consequently, aerosols impact the atmospheric energy budget and temperatures within Earth's atmosphere and at Earth's surface. This scientific definition of aerosol should not be confused with the common usage in association with so-called "aerosol" spray cans and their contents. (See glossary entry for Chlorofluorocarbons.)

Atmosphere

The atmosphere is the gaseous envelope surrounding Earth, which is retained by Earth's gravitational pull. The first 80 km above Earth contains 99% of the total mass of Earth's atmosphere and is generally of a uniform composition (except for a high concentration of ozone, known as the stratospheric ozone layer, at 19 to 50 km). The gases that make up the atmosphere are nitrogen, 78.09%; oxygen, 20.95%; argon, 0.93%; carbon dioxide, 0.04%; and minute traces of neon, helium, methane, krypton, hydrogen, xenon, and ozone as well as trace amounts of water vapor, the distribution of which is highly variable. Earth's atmosphere features distinct layers: the troposphere, the stratosphere, the thermosphere, and the exosphere. The term "free atmosphere" refers to the portions of the atmosphere that lie above the troposphere. (See glossary entries for Stratosphere and Troposphere.)

Biofuels

Biofuels are solid, liquid, or gaseous fuels consisting of, or derived from biomass (plant material, vegetation, or agricultural waste used as energy sources). Biofuels can aid in the mitigation of greenhouse gas emissions by providing carbon neutral alternatives to fossil fuel burning. Burning biofuels does not lead to any net increase in greenhouse gas concentrations because the carbon released when biofuels burn was just recently removed from the atmosphere and has only been stored in plants for a few months. For every gram of CO_2 released when a biofuel is burned, a gram was removed from the atmosphere by photosynthesis just a short time ago. This balance is why biofuels are considered "carbon neutral." Liquid or gaseous biofuels can be used for transport, while solid biofuels can be burned for heat or to generate electric power. Common currently used biofuels, such as ethanol, are derived from maize (corn). Cleaner and more efficient biofuels, such as cellulosic ethanol, derived from switchgrass, are currently under development.

Carbon cycle

The carbon cycle is the sum of the processes, including photosynthesis, decomposition, respiration, weathering, and sedimentation, by which carbon cycles between its major reservoirs: the atmosphere, oceans, living organisms, sediments, and rocks. (See glossary entry for Photosynthesis.)

Carbon dioxide (CO_2) equivalent

Carbon dioxide equivalent expresses the amount of CO_2 that would have the same global warming potential as a given greenhouse gas measured over some defined timeframe, typically one century. It is typically measured in gigatons ("Gt CO_2 eq"). (See glossary entry for Gigaton.)

Chlorofluorocarbons (CFCs)

Chlorofluorocarbons (CFCs) are synthetic compounds consisting of a carbon atom surrounded by some combination of chlorine and fluorine atoms. CFCs are powerful greenhouse gases. However, they are better known for their role in ozone depletion. CFCs were formerly widely used as refrigerants, propellants (in so-called "aerosol" cans), and as cleaning solvents. When their use was implicated in the destruction of the protective stratospheric ozone layer in 1989, the industrial use of CFCs, and other related compounds, was prohibited by the Montreal Protocol. (See glossary entry for Ozone.)

Climate forcing

Changes in the global energy or "radiative" balance between incoming energy from the Sun and outgoing heat from Earth lead to changes in climate. There are a number of mechanisms that can upset this balance, for example fluctuations in Earth's orbit, and changes in the composition of Earth's atmosphere. In recent times, Earth's atmospheric composition has changed as a result of human-generated greenhouse gas emissions. By altering the global energy balance, such mechanisms "force" the climate to change. Consequently, scientists call them climate forcing or radiative forcing mechanisms.

Climate proxy

Climate proxies are indirect sources of climate information from natural archives such as tree rings, ice cores, corals, cave deposits, lake and ocean sediments, tree pollen, and historical records. Information from climate proxies can be used to reconstruct climate for times prior to the establishment of a widespread instrumental atmospheric and oceanic data set. (See glossary entry for Instrumental record).

Cryosphere

The cryosphere is a term for the cold regions of the planet where water persists in its frozen form, i.e., regions covered with glaciers and ice sheets, or with permanently frozen soils. The cryosphere plays different roles within the climate system. The two continental ice sheets of Antarctica and Greenland actively influence the global climate and may also have effects on sea level. Snow and sea ice, with their large areas and relatively small volumes, are connected to key interactions and feedbacks on global scales, including solar reflectivity and ocean circulation. Perennially frozen ground (permafrost) influences soil water content and vegetation over vast regions, and is one of the cryosphere components that is most sensitive to atmospheric warming trends.

Deep time

Distant geologic history, generally before the recent ice-age cycles began 2 million years ago, is often referred to as deep time.

El Niño

El Niño is a climate event in the tropical Pacific ocean and atmosphere wherein the trade winds in the eastern and central tropical Pacific are weaker than usual, there is less upwelling of cold subsurface ocean water in the eastern Pacific,

and relatively warm water spreads out over much of the tropical Pacific ocean surface. During an El Niño event, the warmer tropical Pacific surface ocean waters influence the overlying atmosphere and alter the patterns of the extratropical jet streams of the northern and southern hemisphere and the general circulation of the atmosphere. The altered circulation of the atmosphere leads to changes in temperature and precipitation patterns in many regions across the globe. The name "El Niño" (literally "the boy child") derives from the Spanish term for the Christ Child and originates in the fact that the warming of the ocean waters off the Pacific coast of South America is usually most pronounced around Christmas time. (See glossary entries for ENSO, Jet stream, and La Niña.)

El Niño/Southern Oscillation (ENSO)

The El Niño/Southern Oscillation or ENSO phenomenon is an irregular oscillation in the climate involving interrelated changes in ocean surface temperatures and winds across the equatorial Pacific, which influences seasonal weather patterns around the world. ENSO is associated with alternations between El Niño climate events in certain years and La Niña events in others. (See glossary entries for El Niño and La Niña.)

Energy balance model (EBM)

An energy balance model (EBM) is the simplest type of global climate model. An EBM can be used to determine average global temperature by computing the balance between incoming (solar) and outgoing (terrestrial) radiation. (See glossary entry for General circulation model.)

Fossil fuels

Fossil fuels are hydrocarbon-based energy sources formed over millions of years when the fossilized remains of dead plants and animals are

exposed to heat and high pressure in Earth's crust. Fossil fuels exist in solid form as coal, shales, and methane "clathrate" (methane gas entrapped in a water-ice cage); in liquid form as oil; and in gaseous form as so-called natural gas (mostly methane). Nearly 90% of the world's primary energy production comes from the combustion of fossil fuels. When fossil fuels are burned, greenhouse gases are released into the atmosphere. (See glossary entry for Greenhouse gases.)

Fuel cell technology

A fuel cell can be used to convert fuel energy into electricity in a manner similar to, but distinct from, a battery. In a fuel cell, electricity is generated from the reaction of chemical fuel stored at one end of the cell with an oxidant at the other end of the cell. Unlike in a conventional battery, reactants are consumed during the operation of the fuel cell and must therefore be replaced for continuous electricity production. The primary application of fuel cell technology is in the area of transportation. There are a variety of fuels that can potentially be used in fuel cells. If hydrocarbon-based fuels are used, fuel cells emit only marginally less greenhouse gases than conventional carbon-based energy sources. However, alternatives such as hydrogen cell technology, which are currently being researched, could provide a carbon-free alternative to fossil-fuel based transportation.

General circulation model (GCM)

A general circulation model (GCM) is a three-dimensional numerical model used in global climate prediction and assessment. Unlike simpler energy balance models (EBMs), GCMs can be used to solve for a variety of variables including wind patterns, air pressure, atmospheric humidity, and precipitation patterns. While the most basic GCMs model the behavior of the atmosphere alone, climate modelers often use "coupled" versions of GCMs wherein the atmosphere is allowed to interact with models of the global oceans, the major (Greenland and Antarctic) ice sheets, and terrestrial ecosystems. (See glossary entry for Energy balance model.)

Geothermal energy

Geothermal energy is generated from the heat stored beneath Earth's surface, typically through the use of steam-driven turbines. Often water is injected into the hot subsurface of Earth to generate steam. The use of geothermal energy sources dates back to the early 20th century. Geothermal power provides less than 1% of global energy production, but it is used more consistently in certain regions such as Iceland, and California and Nevada in the United States. Potential exists for more widespread use of this renewable energy source.

Gigaton (Gt)

A gigaton (Gt) is a metric unit of mass, equal to 1 billion metric tons (tonnes). In the context of greenhouse gas emissions, gigatons are commonly used as units for measuring global quantities of carbon dioxide or carbon. (See glossary entry for Carbon dioxide (CO_2) equivalent.)

Global warming potential (GWP)

Global warming potential (GWP) is a measure of how much a given mass of greenhouse gas is estimated to contribute to global warming relative to the same amount of carbon dioxide (see p.28). Since GWP is a measurement of the integrated warming impact of greenhouse gas emissions, it must be calculated and stated for a specific time interval, typically one century.

Greenhouse gases (GHGs)

Greenhouse gases (GHGs) are gases in Earth's atmosphere that absorb longwave radiation including the radiation emitted from Earth's surface (i.e., terrestrial radiation). Because they absorb terrestrial radiation, these gases have a warming influence on Earth's surface (referred to as the "Greenhouse Effect"). Greenhouse gases exist naturally in Earth's atmosphere in the form of water vapor, carbon dioxide, methane, and other trace gases, but atmospheric concentrations of some greenhouse gases such as carbon dioxide and methane are being increased by human activity. This occurs primarily as a result of the burning of fossil fuels, but also through deforestation and agricultural practices. Certain greenhouse gases, such as the CFCs, and the surface ozone found in smog (which is distinct from the natural ozone found in the lower stratosphere), are produced exclusively by human activity.

Hadley circulation/hadley cell

The pattern of rising moist air near the equator and sinking dry air in the subtropics is referred to as the "Hadley Cell" or the "Hadley Circulation" after the 18th-century English amateur scientist George Hadley who first formulated a theory about this atmospheric circulation system. The Hadley Circulation is a key component of the general circulation of the atmosphere; it helps to transport heat from the equatorial region to higher latitudes and is responsible for the trade winds (easterly surface winds) in the tropics (see p.89).

Hydropower

Hydropower is power produced by capturing the kinetic energy of moving water. It is currently the most commonly used source of renewable energy, responsible for just over 6% of global energy production. Hydropower has been used in primitive forms (e.g., for powering gristmills or for irrigation) for many centuries. While obtaining direct mechnical energy from hydropower requires proximity to a moving water source, modern conversion of hydropower to electric power allows long-distance transport of energy, albeit with energy loss (that increases with distance from the source). The availability of hydropower on a regional basis in the future could be affected by shifting patterns of rainfall and runoff associated with climate change.

Ice sheets/Ice Age/ glaciers/glaciation/glacial/ interglacial

Glaciers are huge masses of ice formed from compacted snow. An ice sheet is a mass of glacier ice that covers surrounding terrain and is greater than 50,000 km^2. (The only current ice sheets are in Antarctica and Greenland.) An ice age is a cold period resulting in an expansion of ice sheets and glaciers. This expansion is referred to as glaciation. Ice ages are marked by episodes of extensive glaciation alternating with episodes of relative warmth. The colder periods are called glacials, the warmer periods are referred to as interglacials.

Instrumental record

In the context of climate data, the instrumental record refers to the relatively brief record of direct measurements recorded by instruments such as thermometers, barometers, rainfall gauges, and other devices that measure atmospheric temperature, pressure, wind, humidity, and precipitation, as well as ocean temperature, salinity, water density, and currents. For the atmosphere and ocean surface only, widespread measurements are available as far back as 100 to 150 years. For the free atmosphere and deep ocean, such measurements are generally only available for the past five or six decades. A few isolated instrumental climate records from several centuries back in time are available for regions such as Europe. To obtain climate data from the distant past, climate scientists turn to climate proxy records. (See glossary entry for Climate proxy.)

Intertropical convergence zone (ITCZ)

The Intertropical Convergence Zone (ITCZ) is a belt of low surface pressure that is centered near the equator but migrates north and south within the tropics as the seasons change. The ITCZ is associated with trade winds that converge near the equator, ascending as warm, moisture-laden air currents that rise deep into the upper troposphere in towering cumulous clouds and rainfall-producing thunderstorms. These winds eventually sink in subtropical latitudes but not before their original high moisture content has dissipated. (See glossary entry for Hadley circulation.)

Isotope

Isotopes are atoms of the same element having the same atomic number but different mass numbers. The nuclei of isotopes contain identical numbers of protons but have differing numbers of neutrons. Isotopes of a given element have the same chemical properties but somewhat different physical properties. Some isotopes are radioactive, which makes them useful for dating ancient materials (e.g., carbonaceous materials, rocks, etc.).

Jet stream

The jet stream is a high-speed wind current that lies roughly at the boundary between the troposphere and stratosphere at 8–17 km above Earth's surface. The major jet stream of each hemisphere (referred to as the "polar jet stream") is located at middle/sub-polar latitudes, while a weaker "subtropical jet stream" is found at lower, sub-tropical latitudes in each hemisphere. Both of the jet streams circle the globe as westerly winds (i.e., winds moving from west to east), and, like the ITCZ shift north and south with the seasons. (See glossary entries for Troposphere, Stratosphere, and ITCZ.)

La Niña

In a La Niña event, the trade winds in the eastern and central tropical Pacific are stronger than usual, there is greater upwelling of relatively cold subsurface ocean water in the eastern Pacific, and that cold water spreads out over tropical Pacific ocean surface. During a La Niña event the tropical Pacific ocean and atmosphere are in the opposite state as they are during an El Niño event and the influence on atmospheric circulation and global weather patterns is roughly, though not precisely, the opposite. The term "La Niña" means "the girl child" in Spanish. (See glossary entries for El Niño and ENSO.)

Longwave radiation

Longwave radiation (sometimes referred to as "infrared" radiation) is electromagnetic radiation typically associated with heat or thermal radiation. Atmospheric infrared radiation is monitored to detect trends in the energy exchange between Earth and its atmosphere. These trends provide information on long-term climate changes. Along with solar radiation, longwave radiation is one of the key components of Earth's energy balance studied by climate researchers. (See glossary entry for Solar radiation.)

Megaton (Mt)

A megaton (Mt) is a metric unit of mass, equal to 1 million metric tons (tonnes). There are 1000 megatons in a gigaton. In the context of greenhouse gas emissions, megatons are commonly used as units for measuring global quantities of carbon dioxide or carbon. This usage should not be confused with the distinct usage of the same term to describe the explosive power of a nuclear weapon.

Metric ton

One metric ton (tonne) is 1000 kilograms, roughly equivalent in mass to the Imperial ton (2200 pounds).

Microwave

A microwave is a high-frequency electromagnetic wave; its wavelength is between infrared and short-wave radio wavelengths. Microwaves measurements made by satellites provide one means of monitoring atmospheric temperature changes.

North Atlantic Oscillation (NAO)

The North Atlantic Oscillation (NAO) is a measure of the strength and direction of the predominantly westerly winds that blow across the North Atlantic ocean. The measurement is based on the surface pressure difference between the subpolar and subtropical regions over the North Atlantic ocean. The size of this pressure difference, and the strength and direction of the surface winds, varies from year to year, as first noted by Sir Gilbert Walker in 1932. The NAO can have a profound influence on temperature and precipitation patterns of the North Atlantic and neighboring regions of Europe and North America, particularly during winter in the northern hemisphere.

Organization for Economic Co-operation and Development (OECD)

The Organization for Economic Co-operation and Development (OECD) is an international organization consisting of 30 developed countries that subscribe to a set of Western political and economic principles. The OECD was formed in 1948. It was originally a group of European nations empowered to administer the Marshall Plan for post-World War II European reconstruction. Later, the group was expanded to include non-European Western nations. In 1961, the group formed the Organization for Economic Co-operation and Development.

Ozone

Ozone is a compound of oxygen that is made up of three atoms (the oxygen gas we breathe contains two oxygen atoms). Ozone is an irritating gas, when encountered in surface air pollution. However in the stratosphere, a natural layer of ozone protects life on Earth from harmful ultraviolet radiation from the Sun. CFCs have been implicated in the destruction of this ozone layer. Scientists predict that global warming will lead to a thinner ozone layer, because as the surface temperature rises, the stratosphere will get colder, making natural replenishment of ozone slower and its destruction faster. (See glossary entries for Solar radiation and Chlorofluorocarbons.)

Paleoclimatology

Paleoclimatology is the study of Earth's climate system in prehistoric periods (the paleoclimate), covering the past few centuries to billions of years. Among the methods employed in paleoclimate studies are the simulation of past climates with different continental configurations, ice sheet distributions, and greenhouse gas concentrations using climate models such as EBMs and GCMs, and the reconstruction of past climates from empirical data such as proxy climate records. (See glossary entries for EBM, GCM, and Climate proxy.)

Photosynthesis

Photosynthesis is the process by which green plants and certain other organisms synthesize carbohydrates from carbon dioxide and water using light as an energy source. The common form of photosynthesis releases oxygen as a byproduct. Photosynthesis is part of the carbon cycle and is instrumental in removing CO_2 from the atmosphere. (See glossary entry for Carbon cycle.)

Proxy

See Climate proxy.

Salinity

The technical term for saltiness in water is salinity. Salinity influences the types of organisms that live in a body of water and the kinds of plants that will grow on land fed by groundwater. Salt is difficult to remove from water, so salt content is an important factor in human water use. The salinity differences between different water masses in the ocean is also a key determinant of large-scale ocean currents.

Solar radiation

Solar radiation is the radiation emitted by the Sun, which generates energy from the nuclear fusion reactions in its interior. Most solar radiation is in the form of visible light and some is in the form of ultraviolet and other wavelengths of radiation. (See glossary entry for Ozone.)

Stratosphere

The stratosphere is the layer of Earth's atmosphere that lies above the troposphere, extending from roughly 8–17 km above the surface (lower near the poles and higher near the equator) to about 50 km above the surface. The lower stratosphere contains a natural ozone layer, which absorbs ultraviolet solar radiation, warming the surrounding atmosphere. The warming of the atmosphere with height inhibits vertical air currents, making the stratosphere a highly stable regime of the atmosphere, in contrast to the troposphere that lies below it. (See glossary entry for Ozone.)

Trade winds

Trade winds are the easterly winds (i.e., winds that move from east to west) that are found near Earth's surface in tropical regions. The rising atmospheric currents found within the ITCZ are associated with the convergence of trade winds. The El Niño/Southern Oscillation (ENSO) is associated with a periodic alternation between weakening and strengthening trade winds in the eastern and central tropical Pacific. (See glossary entries for ITCZ and ENSO.)

Troposphere/free troposphere

The troposphere is the lowest layer of Earth's atmosphere, extending from the surface of our planet to between 8 and 17 km (lower near the poles and higher near the equator). The troposphere is sometimes subdivided into a planetary boundary layer, where the atmosphere is in contact with Earth's surface, and the free troposphere, defined as the layer of the troposphere above the planetary boundary layer. The boundary between the troposphere and the stratosphere above it is called the tropopause. The troposphere contains roughly three quarters of Earth's atmosphere by mass. What we normally think of as "weather" takes place almost exclusively within the troposphere. (See glossary entry for Atmosphere.)

Wind shear

Wind shear is a difference in wind speed and direction between slightly different altitudes. Wind shear, among other things, can determine whether or not conditions are favorable for tropical cyclone development.

Index

Note: References in **bold** refer to the Glossary.

Picture Credits

The publisher would like to thank the following for their kind permission to reproduce their photographs:

(Key: a-above; b-below/bottom; c-centre; f-far; l-left; r-right; t-top)

AIMS - Australian Institute of Marine Science: 115. **Alamy Images:** Acbag 158-159; Alaska Stock LLC 1; David Ball 176-177; Peter Bowater 156-157; Jason Bye 163; Ashley Cooper 181cr; Darren Core 148-149; Gary Crabbe 84-85; David R. Frazier Photolibrary, Inc. 112-113; David Sanger Photography 128-129; Andrew Fox 136-137; Dennis Frates 100; Esa Hiltula 28-29; Hoberman Collection UK 149tc; D. Hurst 181cb, 181clb; Jeff Morgan Hay on Wye 120; Darrin Jenkins 160-161; Andre Jenny 181fcr; Huw Jones 72-73, 156bl; Scott Kemper 172-173; Lancashire Images 149; Patrick Lynch 180br; Malcolm Park Wine and Vineyards 46-47; Ron Niebrugge 134-135; NOAA / Michael Dwyer 79bl; Wolfgang Pölzer 114; Robert Harding Picture Library Ltd 124tl; Joern Sackermann 149tl; Sami Sarkis Underwater 192-193; Steve Smith / SuperStock 150-151; Robert Stainforth 55; Ariadne Van Zandbergen 58-59; Visions of America, LLC / Joe Sohm 164-165, 184c; David Wells 170-171; Janusz Wrobel 181tr. **Byrd Polar Research Center - OSU (The Ohio State University, USA):** 58clb; Lonnie G. Thompson 58bl. **Corbis:** 70-71; Abir Abdullah / epa 190-191; Alan Schein Photography 157fbr; Bill Barksdale 157bl; Chip East / Reuters 185br; Alejandro Ernesto / epa 103; Jose Fuste Raga 156fbr; Annie Griffiths Belt 110-111; Gunter Marx Photography 5cl, 108-109; Dewitt Jones 101; Everett Kennedy Brown / epa 157br; NASA 30-31; Rafiqur Rahman / Reuters 125crb; Dick Reed 146bc, 146bl, 146br, 146-147, 147bc, 147bl, 147br; Jim Reed 6-7; Reuters / Beawiharta 90-91; Rickey Rogers / Reuters 123; Alexander Ruesche / epa 156br; Mike Theiss / Ultimate Chase 102; David Turnley 124bl, 130-131; Barbara Walton / epa 26-27; Nik Wheeler 126-127. **Digital Railroad:** Daniel Beltra / Greenpeace 2-3. **DK Images:** David Peart 116-117; Royal Museum of Scotland,

Edinburgh 42tl. **Florida Keys News Bureau:** Andy Newman 62-63. **Michael and Patricia Fogden:** 119tl. **The Galileo Project:** 80bl. **Getty Images:** AFP 186-187; The Image Bank / Joanna McCarthy 194-195; David McNew 49b; National Geographic / Paul Nicklen 118b; Jewel Samad / AFP 187br; Science Faction / David Scharf 97tr; Stone / Ernst Haas 124cla; Stone / Frank Oberle 122. Stone / Lester Lefkowitz 144-145. **Greenpeace:** Ardiles Rante 174-175. **IODP / TAMU:** 40-41. **John Gunn, Earth & Space Research:** 92-93. **John MacNeill Illustration:** 178-179. **Hans Kerp, Forschungsstelle fuer Palaeobotanik, Westfaelische Wilhelms-Universitaet Muenster:** 42cl. **Prof. em. Bruno Messerli / Geographisches Institut - Physische Geographie - Universität Bern:** 58crb. **Moviestore Collection:** 44. **NASA:** 66. NOAA: 56-57. **PA Photos:** AP Photo / Bullit Marquez 18; AP Photo / Karel Prinsloo 58br. **PhotoEdit:** 125tl. **Photolibrary:** Ifa-Bilderteam Gmbh 75; Phototake Inc. / MicroScan MicroScan 42-43. **PunchStock:** Corbis 157fbl; Digital Vision/Peter Adams 32-33 (b/g); Photodisc / Doug Menuez 4cra, 19. RubberBall 181tc. **Science Photo Library:** British Antarctic Survey 32cr, 32tr; Bernhard Edmaier 14bl; Will & Deni Mcintyre 184bl; Susumu Nishinaga 95br; Friedrich Saurer 23bl, 23cl, 23tl, 72tl; SOHO / ESA / NASA 80-81. **Shutterstock:** Tyler Olson 5fcra, 17ca, 23br, 23cr, 23tr, 39fcr, 74cb, 74clb, 74crb, 75clb, 77ca, 107ca, 141, 155, 196-197. **Statoil:** Bitmap / Kim Laland for StatoilHydro 168-169. **Still Pictures:** Roger Braithwaite 5cla, 98-99. **Sun-Sentinel:** South Florida Sun-Sentinel 5ftr, 142-143. Jeffrey Totaro / Esto: 166, 167. **USGS:** 69t, 116tl, 116tr, 138-139. **Werner Berner:** 32crb.

Jacket images: Front: Alamy Images: Panorama Media (Beijing) Ltd. clb; Getty Images: Riser / Jeremy Walker t.

All other images © Dorling Kindersley
For further information see: **www.dkimages.com**

Author Acknowledgements

We would like to thank our colleagues who provided helpful comments on the book at its various stages, including Richard Alley, Klaus Keller, and Jean-Pascal van Ypersele. We would also like to thank our colleagues at Penn State, who have helped to foster a stimulating environment for discourse on the important topic of climate change and its impacts. These include, among many others, Richard Alley, Mike Arthur, Eric Barron, Tim Bralower, Sue Brantley, Bill Brune, Rob Crane, Ken Davis, Bill Easterling, Jenni Evans, Bill Frank, Kate Freeman, Sukyoung Lee, David Pollard, Jim Kasting, Klaus Keller, Ray Najjar, Art Small, Petra Tschakert, Anne Thompson, and Nancy Tuana. Michael Mann would like to acknowledge his colleagues at RealClimate.org for the many keen insights into the science of climate change that they have shared over the years. Lee Kump similarly acknowledges his fellow members of the Canadian Institute for Advanced Research.

The book benefited greatly from the superb editing skills and unflappable patience of Erin Mulligan at Prentice Hall, and the

graphics and design genius of the Dorling Kindersley publishing professionals: Stuart Jackman (for the design and development of the book); Richard Czapnik (for the page layouts); and Sophie Mitchell (for keeping the process on track). We would also like to thank Clive Savage, David McDonald, and Johnny Pau for additional design work; Sue Malyan and Jenny Finch for additional editorial work; and Sue Lightfoot for the index.

Michael Mann dedicates this book to the memory of his brother, Jonathan, and to his daughter, Megan, who will grow up on an Earth whose destiny rests in our hands. He thanks his wife, Lorraine Santy, for her support, and his parents, Larry and Paula Mann for the encouragement they have always provided. Lee Kump thanks his wife, Michelle, and his children, Katie and Sean, for their patience and support during the writing of this book. He dedicates the book to his mother, Patty Kump, for planting the seeds of wonder, and his father, Warren Kump, for his apprenticeship into the world of science.